FROM CHAOS
TO STABILITY

. . . .

T0274562

The New Neuroscience

Ted Abel and Joshua Weiner

SERIES EDITORS

FROM
CHAOS
··· TO ···
STABILITY

How the Brain Invents Our
Conscious Worlds

· · · · ·

Israel Rosenfield and Edward Ziff

Illustrated by Fiammetta Ghedini

UNIVERSITY OF IOWA PRESS · IOWA CITY

University of Iowa Press, Iowa City 52242

Copyright © 2024 by Israel Rosenfield and Edward Ziff

Images copyright © 2024 by Fiammetta Ghedini

uipress.uiowa.edu

ISBN 978-1-60938-989-5 (pbk)

ISBN 978-1-60938-990-1 (ebk)

Printed in the United States of America

Design by April Leidig

Cataloging-in-Publication data is on file
with the Library of Congress.

In memory of Catherine Temerson

CONTENTS

PREFACE

· · · · ·

We are intimately aware of the functions that our brain provides
to us—our ability to perceive our surroundings, to form memories
of these perceptions, and to use this knowledge to make informed
actions and behaviors. And we are beginning to understand how the
brain provides these functions through the physiology of its circuits
and the plasticity of its neurons. Yet although the basic features of
brain function, features of perception and of memory that we experience moment to moment, are thoroughly familiar to us, they remain enigmatic and ultimately raise fundamental questions about
how the brain serves us. We ask: Is there a deep function of the brain
that we can appreciate only if we step back and take a broad view?
This book attempts to answer that question and makes the hypothesis that the basic function of the brain is to simplify a world that
is chaotic and unknowable. The brain creates, or invents, our sensory perceptions, and it presents this simplification to us through
our consciousness so that we may cope with the complexity of the
world and successfully manage our environment in order to survive.
We support this hypothesis with examples drawn from the normal
brain and from the consequences of neurological disease.

The view of the brain that we present was inspired by the thinking of our coauthor, Israel Rosenfield, who, until his passing in Oc-

tober 2022, took consistently fresh views of brain function and made insightful hypotheses. Rosenfield's background was well suited to a creative analysis of the brain. Rosenfield studied mathematics while an undergraduate at New York University and medicine while a medical student at NYU School of Medicine. Later, Rosenfield studied principles of human psychology while a doctoral student in politics at Princeton University, where he wrote a thesis on Sigmund Freud's theory of unconscious motives. He was also a devoted cellist and lectured to students at John Jay College in New York about topics from quantum mechanics to theories of evolution.

Over the years, Rosenfield expressed his thoughts in several highly regarded books, including *The Invention of Memory: A New View of the Brain* (1988), *The Strange, Familiar, and Forgotten: An Anatomy of Consciousness* (1992), and *Freud's Megalomania* (2000), plus numerous articles in the *New York Review of Books*.

This book arose from a fifty-year friendship between Rosenfield and Edward Ziff. Rosenfield and Ziff were graduate students together at Princeton in the 1960s, where Ziff studied RNA structure for his thesis. The friendship was enduring as Rosenfield's career took him to a professorship at the City University of New York, and Ziff's path led to a professorship at NYU School of Medicine, by way of Cambridge, London, and Rockefeller University.

During these times, Ziff mostly researched gene expression, and Rosenfield and Ziff, when ten years out of graduate school, together with the English illustrator Borin van Loon, wrote a provocative, humorous, and highly unusual book about DNA, *DNA for Beginners* (1983), with a second edition, *DNA: A Graphic Guide to the Molecule That Shook the World* (2011). Rosenfield and Ziff also coauthored articles in the *New York Review of Books* about topics in biology, in particular brain function. During this period, Ziff changed his research from gene expression to neuroscience, a change that motivated Ziff and Rosenfield to collaborate again, on this book. This was an unforgettable and exuberant time for these writers, whose endless discussions about biology fueled their writings.

Rosenfield's writings were not restricted to insights about science, and indeed one of his books, *La bombe ou L'histoire véridique de la molaire d'Hitler* (*The Bomb or The True Story of Hitler's Molar*) (2016), is a mischievous fictional update about German advances in constructing an atomic bomb during the Second World War, a graphic novel that was a collaboration with the Italian illustrator Fiammetta Ghedini. Ghedini received a PhD in innovative technology from Scuola Superiore Sant'Anna (Pisa) and University College London in 2010. Since then, she has been combining her scientific background with a lifelong passion for drawing by producing comics and illustrations. Recently, Ghedini set up an agency of science illustrations and comics called RIVA (Research and Innovation through Visual Arts). She often speaks about science communication at public events, and she created the highly original illustrations found in these pages.

At the time of Israel Rosenfield's death, this book was nearly fully drafted, and we have brought the book to completion, motivated by our deep respect for Israel and a desire to make insights about the brain that he inspired available to the public.

The authors wish to express their thanks to Jim McCoy, director of the University of Iowa Press, and to Susan Hill Newton, managing editor, and to Josh Weiner and Ted Able, series editors for UIP's

New Neuroscience series, for their invaluable support and guidance during the preparation of this book for publication. They have our deep gratitude. We also thank Nicole Wayland, Karen A. Copp, and Maya Torrez for expert editing and production efforts that made publication possible.

Fiammetta Ghedini, Paris
Edward Ziff, New York

INTRODUCTION

· · · · ·

Color helps to express light,
not the physical phenomenon,
but the only light that really exists,
that in the artist's brain.

———

HENRI MATISSE

olor is a powerful sensation. Whether the color of a freshly mowed field, a wildflower, or the last rays of sunset, color can catch our attention, inspire us, lift our spirits, or make us melancholy. Color is a creation of the brain, just one of many of the brain's sensory creations, and its presence within our sensory world provides deep clues about the nature and function of the brain. In fact, without a brain, the visual world would be shifty, dirty, and colorless, only electromagnetic waves of many frequencies. We are not aware of this chaotic world because the brain correlates light wave levels in the three frequencies that the human eye is sensitive to and creates from these sensations the colors that we are used to seeing. Objectively, there are no colors in the world. Color is a subjective creation that greatly simplifies and stabilizes the environment, making it possible for the organism to identify people, shapes, and objects, and to move around without bumping into trees and other objects. Hence a chaotic visual world that is ever changing (the shifty, colorless world) can be understood. The brain can make sense of a world with colors and other perceptual creations. Similarly, the sounds that the vocal system produces are as messy and incoherent as the shifty world that confronts our visual system. Again, by correlating various productions of the vocal system, the brain creates the words that we hear.

The world we live in is chaotic and, from a perceptual point of view, unlabeled and disordered. Our brains must find a way to make sense of the chaos by creating our perceptual worlds. Colors and smells do not exist as such; the brain "invents" them to simplify our perceptual environments.

The Enlightenment Challenge to Objectivity

According to traditional thinking, our perception of the external world, received through our senses, is an internal re-creation of an external reality, an inner duplication of the world in which internal perceptions are direct representations of the external environment. However, as ideas that arose during the Enlightenment spread across Europe and North America during the seventeenth and eighteenth centuries, scientists and philosophers presented an opposing view, that our perceptual worlds are not duplications of the external world but instead subjective creations of the brain. Isaac Newton, speaking of rays of light diffracted into a spectrum by a prism, wrote, "For the rays to speak properly are not coloured. In them there is nothing else than a certain power and disposition to stir up a sensation of this or that Colour." Newton, a keen observer of the physical world, continued: "For as Sound in a Bell or musical String, or other sounding Body, is nothing but a trembling Motion, and in the Air nothing but that Motion propagated from the Object . . . so Colours in the Object are nothing but a disposition to reflect this or that sort of rays more copiously than the rest" (*The First Book of Opticks*, 1704, 90).

Newton's observations were at the forefront of a seventeenth-century scientific revolution that led Enlightenment philosophers of the eighteenth century to challenge the notion of an objective worldview provided by the senses. They held that because knowledge of our environment is acquired through our senses, that knowledge must be subjective. In his "Essay on Human Understanding" (1689), John Locke distinguished between primary qualities of an object, such as size, shape, motion, number, and solidity, which he held are independent of the observer and direct extensions of an external reality, and secondary qualities, such as temperature, smell, taste, and sound, which are only known through the sensations they create in the observer and are therefore subjective. Similarly, the Irish philosopher Bishop George Berkeley stated the following in *Three Dialogs Between Hylas and Philonous*:

> You hold that the ideas we perceive by our senses are not real things but images or copies of them. So our knowledge is real only to the extent that our ideas are the true representations of those originals. But as these supposed originals (or real things) are in themselves unknown, we can't know how far our ideas resemble them, or indeed whether they resemble them at all. (1906, 113)

Berkeley further asserted that "light and colors, heat and cold, extension and figures—in a word the things we see and feel—what are they but so many sensations, notions, ideas, or impressions on the sense? And is it possible to separate even in thought any of these from perception?" (*Principles of Human Knowledge*, 1881, 196).

The German philosopher Immanuel Kant argued that all qualities, both the primary and the secondary, are "subjective." "One could, without detracting from the actual existence of outer things, say of a great many of their predicates: they belong not to these things in themselves, but only to their appearances and have no existence of their own outside our representation. . . . To these predicates belong warmth, color, taste, etc." (*Prolegomena to Any Future Metaphysics*, 2004, 40).

Theories of Consciousness

If, as the Enlightenment philosophers asserted, all of our perceptions are subjective, how do these perceptions relate to the objective world, and, most challenging, how do they give rise to consciousness? Modern scientists and philosophers have struggled with these questions and proposed numerous theories, both philosophical and scientific. Most agree that different types of sensory information (visual, auditory, and so on) must be integrated in order to create the single, personal point of view that we experience as consciousness. Indeed, with the global workspace theory of Bernard Baars, consciousness is created by massively parallel processors in the brain that analyze and transform sensory information and other information types. An information exchange system switches the processors back and forth between the conscious and the unconscious states, at any one time, granting to one particular process, access to consciousness. With the global workspace theory, the brain may be compared to a theater where many performers prepare their acts offstage (in the unconscious), and at any one time, only one can take center stage and perform in the spotlight of consciousness.

With the integrated information theory (IIT) of Giulio Tononi and Christof Koch, the essential element of consciousness is the integration of information. With the IIT, consciousness consists of a group of interdependent, information-bearing mechanisms that feed upon and regulate one another. The resulting integration of information is essential to consciousness, and the IIT proposes that consciousness may be found wherever integrated information exists. Consciousness, far from being an exclusively human quality, according to the IIT, also resides in dogs, bats, and other living creatures, and in the extreme, even in certain inanimate objects, if only just a bit, if they possess integrated information.

Anil Seth of the University of Sussex regards consciousness as a controlled hallucination, with an appearance of reality rather than reality itself. Consciousness is capable of predictive processing, creating predictions of the origins of our sensory inputs, predictions that help us navigate the world.

Donald Hoffman of the University of California–Irvine dissociates conscious perception from reality more thoroughly by asserting that "the perceptual systems with which we have been endowed by natural selection are a species-specific interface that allows us to interact adaptively and successfully with objective reality, while remaining blissfully ignorant of the complexity of that objective reality" ("The Interface Theory of Perception," *Current Directions in*

Psychological Science 25, 2016). According to Hoffman, our perceptions do not approximate the objective world but instead provide a simplified user interface to that world, much as icons on a computer desktop provide a functional interface with the computer's software and electronics.

Despite the insights provided by these and other theories, we are far from understanding consciousness, and in particular it is difficult to account for how the brain transforms the objective features of the physical world into the subjective perceptions we experience through consciousness. This challenge led the Australian philosopher David Chalmers to pose what he called "the hard problem of consciousness." How can a physical structure such as the brain give rise to the personal experiences that we associate with consciousness, such as the taste of an apple or the experience of joy? Philosophers have named these personal experiences "qualia" (singular, "quale"). The philosopher Thomas Nagel has described qualia as "what it is like" to undergo a particular experience, such as what it is like to feel a twinge of pain or to see the color green. Qualia may encompass various types of personal experience, such as "what it is like" to experience bodily sensations (feeling warmth) or feeling passions or emotions (delight) or moods (elation).

The Berkeley philosopher John Searle has posed the hard problem from the perspective of a neuroscientist: "The problem, in its crudest terms, is this: How exactly do brain processes cause conscious states and how exactly are those states realized in brain structures?"

In an attempt to answer this question, Francis Crick, codiscoverer of the double helical structure of DNA, gathered, during the later stages of his career, scientific insights into the problem of consciousness in his book *The Astonishing Hypothesis: The Scientific Search for the Soul* (1994). Crick's book is a compendium of physiological and anatomical brain phenomena related to consciousness, ranging from visual illusions to functions of neural networks. The book provides penetrating and forceful support for Crick's "astonishing hypothesis" that "you, your joys and your sorrows, your memories and your ambitions, your sense of personal identity and free will, are in fact no more than the behavior of a vast assembly of nerve cells and their associated molecules" (3). Yet, despite a great amount of thought and experimentation, the mechanism by which the brain transforms information about the physical world into consciousness remains unknown.

Our Thesis:
The Brain Invents Our Sensory World

If the origin of consciousness is elusive, what can we say about how the brain serves us? In our view of brain function, the brain creates our perceptual environment to simplify a world that is chaotic and unlabeled. We go beyond Enlightenment views and hold that the entirety of our conscious world is fabricated by the brain, and that the creation of a coherent environment out of chaotic stimuli is one of the brain's primary activities. There are no colors in nature, only electromagnetic radiation of varying wavelengths. Without an animal brain to interpret them, words and sentences would be just a jumble of sounds, whistles, grunts, and silences, and our wines and filet mignons tasteless and odorless. Brains create something that is not there, and in doing so they help us understand and manipulate our environment.

We show that memory is not an orderly archive or fixed representation of things past but a dynamic and malleable function of the brain that is relational. Memory is formed as the brain creates relations, such as the spatial relationship of self to things around us or the temporal relationship of events. Memory may be updated to increase its utility to the individual, and it evolves with time and reflects our subjective access to past histories of self. Hence memory is the result of a process of reconstruction in which many fragments that are stored in the brain can be reused in a coherent fashion at a particular moment. The original perception is not precisely reproduced.

We hold that malfunctions arise when a brain is unable to simplify. In this book, we describe disorders such as Capgras syndrome, a neurological disorder that can disrupt our recognition of relationships with family members, making us deny that a husband or brother is a relative and assert he is really an impostor. And other brain deficits that impair our ability to correlate letters with words and integers with numbers, as with the illness of Oscar, a patient of the French neurologist Joseph Jules Dejerine, who could see and copy letters with great difficulty but could neither read nor make sense of the very letters he had copied. And the Canadian novelist Howard Engel, who could not read paragraphs of text he had just written. And people with prosopagnosia, who lack the capacity to recognize faces, even those of close friends.

We discuss how artists have always known, intuitively, that the brain makes possible the creation of our visual worlds, such as when artists create the illusion of faces, objects, and scenes on a flat surface, and when Pablo Picasso, Marcel Duchamp, and Georges

Braque represent time, space, and motion through Cubism, and when the artist Chuck Close grids faces onto paintings in two-dimensional arrays to overcome prosopagnosia or face blindness. We also consider how the authors Marcel Proust, Virginia Woolf, and others use literary style to represent the dynamic and ever-changing properties of memory. We go on to place this understanding of brain function and consciousness in the context of the evolution of other scientific fields, including the evolution of classical Newtonian physics into quantum mechanisms and the evolution of genetics from the Darwinian and Mendelian views into the modern understanding of genetics, which incorporates split genes and epigenetics.

Our perceptions of the world, as experienced through consciousness, are subjective, and philosophers and scientists have presented theories for the origin of these perceptions. We hypothesize that because our world is chaotic and unlabeled, the brain invents or creates the sensory features of our environment, such as color, sound, and taste. The brain also creates spatial and temporal relationships that form the basis of memory. These creations of the brain support our survival by simplifying our environment so that we can understand and manipulate it.

In order to place consciousness in a scientific perspective for the nonscientist and to relate brain function to philosophical questions such as the hard problem, we must acquaint ourselves with the fundamentals of how the brain operates. Therefore, for the reader new to this field, in the appendix we also provide basic overviews of brain anatomy and the brain's major constituent cells, the neurons, and review mechanisms used by the brain for transmitting signals during the processing of sensory information.

CHAPTER 1

· · · · ·

The Brain and Perception

Snowstorms can produce dangerous whiteouts—the inability to distinguish the horizon, objects, and even people a few feet away. On roads, drivers become disoriented and suddenly unable to detect vehicles right in front of them. In a related phenomenon, individuals who are put in sensory deprivation tanks lose consciousness and begin to hallucinate. Individuals deprived of normal sensory input for many hours can develop permanent brain damage.

In other words, conscious perception requires constantly changing sensory input. We cannot perceive static images. Our eyes are constantly moving and this movement is essential for us to perceive images or objects.

Information received by the sensory organs, the eye for visual information, the ear for auditory, the nasal passage for olfactory, and the skin for tactile information, travels in the form of nerve impulses to different regions of the brain within the cerebral sensory cortex, with one region dedicated to each sense. Specialized features of the information are extracted, such as the ratio of intensities of different light wavelengths to establish color, or the direction of motion of an object, or the relative levels of odorant molecules for aroma, or frequencies of airwaves for sound. These bits of sensory information are delivered to another brain region, the association cortex, where the different types of sensory information are assembled, along with other sorts of information, such as about pleasure and reward, or fear or states of emotion to create or invent a simplified, unitary, and coherent representation of our sensory environment. This assembled information is used for a number of purposes, including guiding our behavior and abstract thinking.

Thus, the brain transforms the ingredients of our sensory inputs into something new, as if it were baking a cake, and the brain's invention or creation of our sensory world might be viewed as the brain "baking cakes." When baking a cake, you transform the ingredients into something new—for example, with the brain, colors.

Perception of Color

We have asserted that colors do not exist in the objective world. An individual dressed in a red costume, sitting in a room painted red, will seem without specific color, not red. The brain creates colors when there is more than one frequency of light in the visual field. Colors are created when there are at least two frequencies of light.

In the nineteenth century, the English physicist Thomas Young proposed that the variety of colors that one can see could be accounted for if one postulated a limited number of color receptors in the eye. He proposed three: one for red, one for green, and one for blue. He also proposed that the color of each point in a field of view was determined by the relative responses of the three-color

receptors in that field, which in turn depend on the amount of light in each of the wavelengths: long wavelengths for red, medium wavelengths for green, and short wavelengths for blue, with other colors falling between these.

In 1964, Edward MacNichol Jr., at Johns Hopkins University, and George Wald, at Harvard, discovered that there are indeed three types of color receptor pigments in the retina. This confirmed Young's proposal of pigments in the red, green, and blue wavelength ranges, each associated with a different receptor. (The particular pigment associated with a receptor determines the wavelength response of that receptor.) But the assumption that each color receptor was merely measuring the intensity of light in its particular wavelength failed to account for color constancy: the fact that a red object always looks red whether viewed in daylight or with artificial light. It failed also to explain why a uniformly colored environment without edges appears colorless. (For example, airplane pilots experience "graying out" when peering into a cloudless sky.)

It was in the late 1950s that Edwin Land, the inventor of the Polaroid Camera, demonstrated to the American National Academy of Sciences how color could be created from black-and-white images. Land had prepared two black-and-white photographs of a redheaded woman in a green dress. One photograph had been taken with a red filter and the other with a green filter. The photos differed in their black and white densities. In one photograph the dress was darker than in the other, the hair darker, and so on. But they were identical in form and, of course, both were without any color. The two photos were then placed in projectors and the images were superimposed on a screen. A red filter was placed in the projector

with the image that had been taken using a red filter. The other projector had only white light.

Since the photos were in black and white, and the only color being added to the scene was red, one would have expected a reddish-pink image to have appeared on the screen. What, in fact, appeared was the redheaded woman in a green dress. The colors of the original scene had been reproduced, and there was no way the audience could eliminate them. Indeed, any multicolored scene photographed in black and white using first a red and then a green filter, and then projected so that the images are superimposed, one image with a red tint and the other with a white tint, will produce a fully colored scene. The brain simplifies the sensory input by creating something that doesn't exist—colors.

When Land developed an instant color photography camera, two-color systems did not prove adequate, and the actual Polaroid Land instant color cameras employed a three-color system, as is the case with the human retina.

While the three different receptors in the retinas of human beings respond to long-, middle-, and short-wavelength light, birds have at least four different receptors and hence four different frequencies of light to which they are sensitive. Birds see more colors than humans. And one might ask why? What is the evolutionary advantage of seeing more colors if you are a bird? Since birds fly, they need more precision when trying to land on chimneys or when avoiding other birds when they are flying. Humans, in contrast, used technology to develop radar and airplanes.

In 2007, Jeremy Nathans and colleagues working at Johns Hopkins University showed that an increase in the number of photoreceptor types with different photopigments in the retina, a change that enhances comparison of light intensities by the brain, also enhances the perception of color. Unlike birds and humans, most animals, including rats and mice, have only two photoreceptors, with two photopigments in their retinal cones, a short-wavelength light-sensitive receptor (sensitive to blue) and a medium-wave light-sensitive photoreceptor (sensitive to green). Because humans and certain other primates have a third, long-wavelength light-sensitive receptor (sensitive to red), they can see a larger range of colors than rats or mice. Nathans and colleagues used genetic engineering to insert the gene encoding the human long-wave light-sensitive photoreceptor into mice. Behavioral tests showed that this insertion increased mouse sensitivity to long-wave (red) light and also increased the number of colors the mice could discriminate when compared to the genetically unmodified mice. When the number of photoreceptor types is increased, the brain can make a larger number of comparisons of light intensities at different wavelengths, allowing the brain to discriminate, perceive, or create a larger number of colors.

Given this effect of adding a photoreceptor, it is not surprising that loss of a receptor through mutation of its gene will decrease color discrimination. The brain invents fewer colors for such a person. This situation is found with people (mostly men) who have mutations in the gene for either the long-wavelength or the medium-wavelength light-sensitive receptor, which causes red-green color blindness. Color-blind people have trouble perceiving different shades of red, green, and yellow, showing that a decrease in ability to compare the light intensities decreases the number of colors our brain can invent.

While vision employs photoreceptors that respond to photons (units of light), other senses such as the olfactory, auditory, and somatosensory or tactile sense use receptors that respond to the relevant physical properties of the environment.

Vision by Frogs, Cats, and Flies

During the 1950s, neurophysiologists investigating the mechanisms of vision discovered neurons in the visual cortex that are activated by specific types of stimuli. It soon became clear that different organisms (humans, flies, frogs) perceive through vision different aspects of their environment that are suited to their particular needs. Within the frog's brain, they found detectors that fired whenever a moving convex object appeared in a specific part of the frog's visual field. If the object failed to move, or if it was the wrong shape, the neuron would not fire: hungry frogs would not jump at dead flies hanging on strings, but they would if the string was jiggled. In their studies of cat and monkey visual cortices, David Hubel and Torsten Wiesel, working at Harvard University, found specific neurons that were sensitive to lines and bars with specific horizontal, vertical, and oblique orientations. The visual cortex apparently responded to particular features such as lines and bars in the physical environment.

Studies on animal vision in the 1950s (the first major study was "What the Frog's Eye Tells the Frog's Brain," published by MIT researchers) suggested that vision served different purposes in different animals. The frog's visual world is very different from the fly or human visual worlds. Each species is seeing what is necessary and

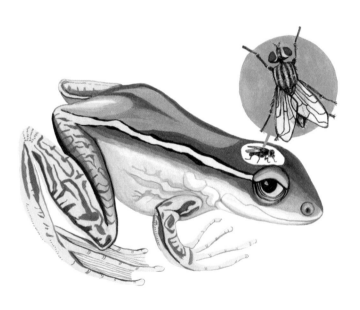

advantageous for its survival. The neurons in the frog's eye respond to convex moving objects that resemble flies, the principal element of the frog's diet. Any such object will cause the frog to jump. A convex stationary object or a dead fly on a string will not excite the retinal cells of the frog, and even a starving frog will not budge. Jiggle the strings attached to the dead fly or convex object and the frog jumps. The frog's visual system responds to stimuli other than moving convex objects, but its range is very limited and directly related to survival of the frog.

A fly's visual systems are concerned with landing, taking off, and smooth flight. An image of an exploding surface (a surface that suddenly increases in size as one rapidly approaches it and therefore explodes) on the retina of the fly causes it to prepare for landing. The fly extends its legs and turns off its wing power as the exploding surface appears in its visual field. In-flight mechanisms, on the other hand, will guide the fly toward other small flying objects. The visual system of the fly is relatively simple, and what the fly sees has little to do with what the frog sees. What is a visual illusion for one species is not necessarily an illusion for another. A fly may confuse a tiny flying insect with an elephant a hundred yards away, but humans would make no mistake.

Artificial Intelligence

As visual processing in animals and humans became better understood, scientists in the field of artificial intelligence decided it should be easy to build seeing machines that could identify and manipulate objects by matching electronically registered shapes with images stored in the computer's memory. This, however, proved considerably more difficult than they had anticipated, in part because much of what we see has nothing to do with the shapes and locations of physical objects—for example, shadows, variations in illumination, dust, or different textures. Which features are important for seeing an object and which can be ignored? Computer scientists found that a seeing robot needed an enormous memory stuffed with photos, drawings, and three-dimensional reproductions of grandmas, teddy bears, bugs, and whatever else the robot might encounter in its preassigned tasks. They tried to simplify the problem by restricting visual scenes to minute worlds of toy blocks and office desks, and they concentrated on writing programs that could effectively and rapidly search computer memories for images that matched those in the robot's eye. Eventually, some of these programs worked well, specifically ones that borrowed strategies from the brain. By incorporating into the software "Deep Learning" computational methods, which mimic aspects of brain function, the ability to identify objects increased greatly. And these advances in computer recognition of images have made possible self-driving cars, which must be capable of interpreting rapidly changing images of roadways and the driving situations they present. But ultimately, artificial intelligence has failed to reveal the deep inner workings of the brain. The nature of brain function had not been recognized: that brains evolved to create a simplified sensory world and that colors, smells, and sounds as such do not exist but are creations of the brain. The brain invents color by comparing intensities of light falling on the retina at different wavelengths. Other sensory perceptions are similarly created. This hypothesis was supported by the Land experiment, in which

full-color images were perceived when viewing black-and-white photo transparences illuminated by monochromatic light. Vision by different organisms (fly, frog) is not a simple representation of the external world but a selective presentation of particular visual aspects depending on the organism's particular need. The invention of a simplified sensory world enables us to navigate a perceptual environment that is chaotic and ever changing.

CHAPTER 2

· · · · ·

Perceptual Processing

How does the brain create a simplified sensory world from a chaotic and unknowable environment? The brain cannot respond to the physical environment directly. Perceptual features of the environment—photons for vision, airwaves for sound, odorant molecules for smell—must first interact with specialized receptor proteins that are mounted on the outer membranes of neurons in the sensory organs and exposed to the outside world. Activation of these receptors generates nerve impulses that are intelligible to the brain. These impulses encode information about the environment, which is transmitted to the various cortical sensory regions for processing. Processing of these diverse forms of information in the cortex extracts features of the environment, in a simplified form, which are assembled to create a coherent view of our world that is presented to us through our consciousness.

Visual information reaches us in the form of photons (particles or quanta of light) that enter the eye and fall upon cells in the retina called rods and cones, which have receptors that induce nerve impulses that travel to the visual cortex. Rod cells convey information about size, shape, and brightness of visual images while cones convey information about wavelength of light, which establishes color by relating intensities at different wavelengths. Individual visual cortex neurons respond to particular patterns of light falling on the retina, such as edges, which are positions of high contrast, rather than to an actual image of a section of an object. After initial processing in the primary visual cortex, neuron activity representing this basic evaluated information passes to the other, specialized visual cortex subregions, where finer forms of processing take place that enable us to perceive and identify complex visual patterns, such as those created by a particular object (a stapler) or, most challenging, by a particular human face.

Odors arise from mixtures of small airborne molecules, volatile chemicals that enter the nasal cavity, where they stimulate olfactory receptors embedded in the outer membranes of olfactory neurons. There are hundreds of different olfactory receptor types, each responding to (binding to and thus detecting) one or a small number of volatile chemicals. The particular pattern of activation of the different receptors that respond to an odor is transmitted to the olfactory cortex, where the pattern serves as a signature for a particular source of odor, such as a perfume or an aromatic food.

In the perception of sound, objects vibrating in the atmosphere move air molecules to create airwaves of different frequencies and intensities that enter the ear, where an inner ear structure, the cochlea, converts them to nerve impulses. The impulses exit the ear through the cochlear nerve, are sorted into different nerve fibers according to sound frequency, and travel to the auditory cortex. The auditory cortex processes the information to discriminate

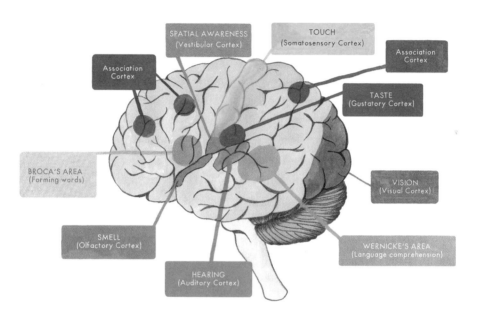

SPATIAL AWARENESS
(Vestibular Cortex)

TOUCH
(Somatosensory Cortex)

Association
Cortex

Association
Cortex

TASTE
(Gustatory Cortex)

BROCA'S AREA
(Forming words)

VISION
(Visual Cortex)

SMELL
(Olfactory Cortex)

WERNICKE'S AREA
(Language comprehension)

HEARING
(Auditory Cortex)

"sound objects," which are akin to the sounds produced by the different instruments in an orchestra, and other sound properties, such as direction.

Pathways of Processing of Sensory Information

After the various physical features of our environment activate sensory receptors and their activities are converted to nerve impulses, the resulting bits of information do not travel directly from the sense organs to the sensory cortex (olfaction, the sense of smell, is an exception). Instead, they pass through a way station called the thalamus, which is a group of nuclei, or bundles of neurons, located in the middle of the brain. There is one thalamic nucleus for each sense except olfaction. Upon arrival in the thalamus, each type of sensory information is processed in one of the nuclei and then transmitted to the relevant region of sensory cortex for further processing. Other sorts of information, information not arising from the physical environment, such as information about emotion, motor activity, pain, and certain cognitive functions, also passes through the thalamus. Because multiple forms of information converge on the thalamus, the thalamus has been proposed to play an important role in consciousness.

There is another "twist" to the path that sensory information takes to the cortex. The left hemisphere of the brain controls actions on the right side of the body and also receives sensory information from the right side, while the right brain hemisphere controls functions of the left side of the body and receives sensory information from the left side, resulting in a "contralateral" organization of functions. Thus, information from the right visual field (which is registered by portions of both eyes) is processed in the visual cortex of the left cortical hemisphere, and information from the left visual field is processed in the right visual cortex. Motor processing is also contralateral, so the motor cortex in the left hemisphere controls the right-side arm and leg, and injury to the right side of the brain causes a deficit on the left side of the body.

Each of the two hemispheres of the cerebral cortex, the right and the left, has a motor cortex region and a sensory cortex region. Within each hemisphere, the sensory and motor cortex regions face one another, separated by a groove called the central sulcus. The motor cortex plans and executes voluntary movements and lies on the anterior (toward the front of the brain) side of the sulcus, while the somatosensory cortex processes tactile information and lies on the posterior (toward the back of the brain) side of the sulcus.

Both the motor and sensory cortices display a striking anatomical organization of function. Stimulation of different positions on the motor cortex controls muscles that move different parts of the body. Stimulation of the uppermost part of the motor cortex makes the knee contract, while stimulation of the motor cortex at the side of the brain causes the tongue to move. Stimulation of adjacent positions creates a pattern of movement that resembles the anatomic arrangement of the human body, and this functional pattern is called the motor homunculus (*homunculus* is Latin for "little man"). The same arrangement is true for tactile sensation. Stimulation of different points on the somatosensory cortex produces a related anatomical pattern of bodily tactile sensations called the sensory homunculus. Body regions, such as the face, that are more responsive

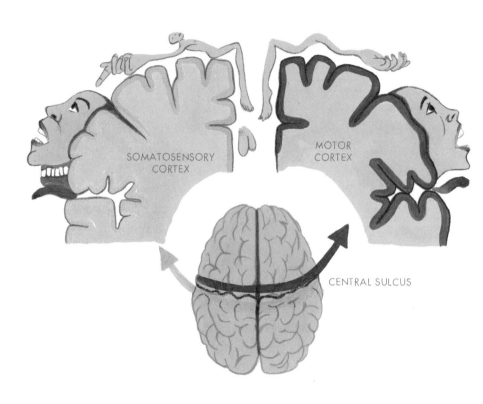

SOMATOSENSORY CORTEX

MOTOR CORTEX

CENTRAL SULCUS

to tactile stimulation have a greater representation in the sensory homunculus. Keeping with the brain's contralateral organization, the homunculi in the right hemisphere correspond to functions on the left side of the body, and vice versa. This anatomical organization is often cited as evidence for cerebral localization of function.

Association Cortices

After sensory information is processed in the respective primary processing areas, it moves to the association cortices, where it is integrated with information from emotional, spatial, and fear-related modalities together with information from the motor cortex, thalamus, and brain stem. It is here that abstract mental functions also take place. Notably in the evolution of the primates including man, the association cortices increased in size relative to the sensory and motor cortices.

Localization of a Simple Behavior to the Spinal Cord

In the search for brain locations that provide other specific functions, researchers surgically destroyed discrete regions of animal brains and determined which functions were lost. Based on this approach, the English neurologist John Hughlings Jackson, a pioneer in this area, postulated that the brain is a sensory motor machine: sensory stimulation could produce motor activity, such as a reflex, which is an automatic action like sneezing that takes place without conscious thought. Some suggested that all mental activity was but a series of reflexes, in particular the English neurophysiologist Charles Sherrington, who proposed that reflexes were the building blocks of animal and human behavior. Stereotyped behavior was produced by chains of reflexes: each reflex became the sensory input for the next one in the chain. Indeed, Sherrington and his coworker, Thomas Graham Brown, could demonstrate normal walking movements in a cat whose spinal cord was severed from the brain, consistent with Sherrington's claim that reflexes formed the basis of behavior.

Localization of Language Functions within the Cerebral Cortex

Other researchers who studied more complex forms of behavior, such as the formation of language, noticed that certain clinical manifestations were associated with lesions in relatively localized areas of the brain. Lesions of this sort occur when the blood supply is interrupted, depriving a brain region of oxygen—that is, when a stroke occurs. Without oxygen, neurons can't produce the energy necessary for their survival and they perish, producing a lesion that often impairs a specific brain function whose nature depends on the lesion's location.

The French physician Pierre Paul Broca proposed that a patient of his with speech impairment at the Bicêtre Hospital in Paris be used as a test case for localization of function. Broca's patient had lost his speech twenty-one years earlier and was known as "Tan" because this was the one word he could utter. (His real name was Leborgne—the French slang for "one-eyed.") Almost too conveniently, Tan was admitted to Broca's surgical service and a few days later died of a gangrenous leg, on April 16, 1861. On April 17, Broca demonstrated Tan's brain to the Société d'Anthropologie. There was a well-circumscribed area in the left hemisphere (today known as Broca's area) that was badly damaged, the consequence of a stroke.

Broca soon found that patients only had speech defects when the damaged area he had identified was on the left side of the brain. Right-sided damage had no effect on speech. Broca's patients were, in general, unable to articulate themselves, but they had no trouble understanding those who addressed them.

The case for localization of function was strengthened in 1874 when Broca's contemporary, the German physician Karl Wernicke, discovered that damage to another area in the left hemisphere (today known as Wernicke's area) resulted in grammatically correct but roundabout speech. His patients had difficulty finding the right words. Unlike Broca's patients, Wernicke's had poor comprehension of speech. Wernicke's work suggested that not only was there localization of function but different aspects of a given function were localized in different parts of the brain.

The way in which functions are broken down in the brain can provide us with clues as to just what the brain is doing, but more than one hundred years after Wernicke, we are still far from understanding why the brain processes information the way it does, though we are beginning to have some ideas. From a clinical point of view, however, the breakdown of a specific function subunit is often pinpointed to a specific lesion in the brain.

We now appreciate that many other functions are localized: fear to the amygdala, reward to the striatum, and many cognitive functions plus working memory and inhibition of impulsivity to the prefrontal cortex. As noted previously, specific forms of sensory processing are localized within specific regions of the sensory cortex and voluntary motor activity in the motor cortex.

Split Brains

Understanding localization of brain function took an unexpected turn through studies of the two hemispheres of the brain and a thick band of nerve fibers called the corpus callosum that connects them. The corpus callosum had been recognized since the early twentieth century to integrate the functions of the two hemispheres. In the late 1950s, surgeons cut the connected fibers in a group of patients with epilepsy in an attempt to control the spread of epileptic seizures. The operations were relatively successful and, at first, it was noted that the sectioning of the corpus callosum had no obvious effect on the mental functioning of the patients. But then, in the 1960s, Michael Gazzaniga, a student of Roger Sperry at Cal Tech and later professor at the University of California, found that these patients manifested subtle defects in the performance of various tasks. Not only was the right hemisphere, or so-called minor hemisphere, capable of a variety of difficult mental tasks, but it also was able to understand both written and spoken language despite the dominant role of the left hemisphere. These studies represented a dramatic confirmation of localization of function, while at the same time raising new questions about the nature of consciousness.

Sperry and Gazzaniga used an experimental setup in which visual cues were given to either the left or right hemisphere of the commissurotomized patient—also called a "split-brain" patient.

As discussed previously, visual pathways from the retina to the brain cross one another, so what is seen in the right visual field is processed by the left hemisphere of the brain and vice versa. A picture of a cat held directly in front of the right eye stimulates exclusively the visual field of the right retina and would be seen exclusively by the left hemisphere (and vice versa). When the patient was asked to identify what he had seen in the right visual field, he had no difficulty saying that he had seen a cat. Since the information was going to the left hemisphere, where the speech centers are located, nobody was very surprised by this result. However, when the same image was presented in the left visual field through the left eye—now going exclusively to the right hemisphere—the patient would deny having seen anything. Nothing had been seen by the left hemisphere, which contains the speech centers, because the main connection between left and right hemispheres had been severed. Nonetheless, when the patient was presented with several pictures, among them that of a cat, he had no difficulty in identifying the cat (using his left hand, responsive to the right motor cortex) as the object that had been seen by the right hemisphere. While the right hemisphere was not capable of *verbalizing* what it had seen, it was capable of *seeing* and *remembering*. It could even understand written words, such as "cat" or "container of liquids." Sentences, however, were poorly understood. Ultimately, it became clear that in commissurotomized patients, each hemisphere was operating independently. Frequently the left hemisphere would attempt to interfere with the performance of tasks initiated by the right hemisphere and would even make comments ("Now, why did I do that?"), though it was totally ignorant of the instructions that had been given to the right side.

The right hemisphere, however, proved to be more than a mute version of the left. Each hemisphere, it soon became apparent, has its own area of specialized mental activities. While one should, nonetheless, be careful about exaggerating the findings, in general, right hemisphere specialization was shown to include understanding and manipulating geometric figures, identifying faces, discriminating musical chords, and making mental spatial transformations. The left hemisphere functions included language skills and mathematical skills.

Since emotional information appears to be processed, in part, in the brain stem and is then passed to both hemispheres (where there are different "emotion" centers), it was not affected by the commissurotomy. Gazzaniga describes a patient who was shown the word "kiss" to the right hemisphere and was told to perform the command. Though the left hemisphere was ignorant of what it was being told to do, the patient said, "Hey, no way, no way. You've got to be kidding." But the patient was unable to say what he was not going to do. When the same word was presented to the left hemisphere, the patient repeated his refusal to kiss the examiners, though he was now able to explain what it was that he was refusing to do. In other words, the emotional reaction of the right hemisphere was correctly sensed by the left. The reasons for that particular reaction, however, remained inexplicable to the left side of the brain.

The two hemispheres can work against each other, depending on the particular information presented to each half. If an image of a face that is half man, half woman is shown to a commissurotomized patient and his or her gaze is fixated at a point in the middle of the image, should the left hemisphere see the man's face it will verbalize that the face is that of a man. But given a set of drawings of whole faces, the right hemisphere will direct the individual's hand to point to a woman's face. (The right hemisphere has seen the female side of the drawing.) Though each hemisphere is seeing just half a face, it will process the information into a consistent whole. Depending on the assigned task—verbalization about the face as opposed to manual identification of the face—one or the other hemisphere will control the individual's actions.

For many years, scientists were puzzled by the fact that visual information is duplicated in the right and left hemispheres. That can now be explained—as we have seen in the commissurotomized patients—by the assumption that each hemisphere is processing the information presented to it differently (using it for different purposes, such as verbal analysis as opposed to geometric analysis of the same image).

These findings led to a considerable discussion concerning the state of the consciousness of the right hemisphere. Since it can't verbalize, is it really "conscious"? One attribute of consciousness that is believed to have developed late in evolution is self-consciousness, or self-awareness. Although humans are self-aware, dogs, for example, show little interest in their reflection in a mirror, or their image in a photo. Studies have shown that, using only the right hemisphere, one can easily recognize oneself in a photo or a mirror. Though not verbalized, photos of close friends and relatives give rise to emotional displays, including that of a sense of humor. So while the left hemisphere can talk about the individual (and even arrogantly criticize the activity of the right hemisphere, though it is ignorant of the information presented to it), the right hemisphere has an equally well-developed sense of itself and its relation with other persons.

Overview

The brain invents or creates our sensory world starting with information types that are highly diverse and for the most part unrelated to one another in physical nature: air vibrations (sound), electromagnetic waves (photons for vision), and airborne molecules (odorants). To make sense of such a complex and incoherent group of properties, the brain must first convert them into a form that is intelligible to the brain—neuron electrophysical activity (nerve impulses). Each property of the environment has its own set of sensory receptors dedicated to this conversion.

Once converted to nerve impulses, specialized information features such as light intensity, sound frequency, or the chemical identity of volatile odorants are extracted in the sensory cortex, united in the association cortex, and then interrelated with other information such as about emotion or pain. The result is the unified, coherent, and simplified representation of our environment that we perceive in our consciousness.

The brain uses this information to select and create complex behaviors, such as generation and interpretation of speech, behaviors that are also formed in specific cortical areas. Studies of patients with surgically disconnected cerebral hemispheres (with split brains) have helped to discriminate the origins of such functions within the two hemispheres. Abstract thinking appears to rely on the association cortex while simple reflexive behaviors arise in the spinal cord.

CHAPTER 3

·····

The Relational Nature of Memory

To make use of the information that the brain extracts from our sensory perceptions, the brain creates a record of it in the form of memory. But how is this information organized and what is recorded?

What Is Stored in Memory?

In a remarkable book published in 1932, *Remembering: A Study in Experimental and Social Psychology,* the English psychologist Frederic C. Bartlett wrote:

> Remembering is not the re-excitation of innumerable fixed, lifeless and fragmentary traces. It is an imaginative reconstruction, or construction, built out of the relation of our attitude towards a whole active mass of organised past reactions or experience, and to a little outstanding detail which commonly appears in image or in language form. It is thus hardly ever really exact, even in the most rudimentary cases of rote recapitulation, and it is not at all important that it should be so. (218)

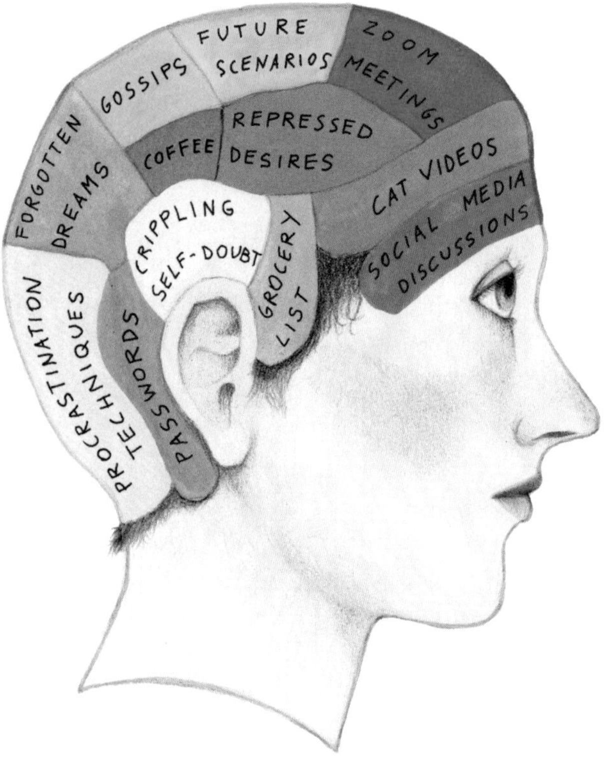
FUTURE SCENARIOS
ZOOM MEETINGS
GOSSIPS
REPRESSED DESIRES
COFFEE
FORGOTTEN DREAMS
CAT VIDEOS
SOCIAL MEDIA DISCUSSIONS
CRIPPLING SELF-DOUBT
GROCERY LIST
PROCRASTINATION TECHNIQUES
PASSWORDS

While remembering, according to Bartlett, is an active reconstruction that is hardly ever exact, for a long time many have held a contrary view and have assumed that our recall of people, places, and things can be accurate. They are accurate because images of them have been imprinted and stored in our brains, and though we may not be conscious of them, these images are the basis of recognition, thought, and action. In this view, we recognize people and things because our brains match what we hear and see with stored images that have been acquired. Why was this notion widely accepted?

By the end of the nineteenth century, neurological discoveries lent considerable credence to the generally accepted view of permanent fixed memories. Paul Broca had discovered the speech production center, and Karl Wernicke had discovered the center for understanding speech, and motor and sensory areas were found. Neurologists concluded that the brain consists of a collection of highly specialized functional regions that control speech, movement, vision, and so on. Closely related to this view was the view that memory, too, was divided into many specialized subunits with memory centers for "visual word images" and "auditory word images." Failure to recognize or recall was explained as the loss of a specific memory image or center.

In the 1930s, the American Canadian neurosurgeon Wilder Penfield made a series of spectacular discoveries that reinforced the theory of localization of function and permanent memories. Penfield discovered that electrical stimulation of certain areas of the brain in conscious patients elicited recollections of "forgotten" experiences—what he called flashbacks. Believers in localization of function and permanent memory traces could not have wished for more compelling evidence. The memories elicited were fragmentary, and patients only had a sensation of having a "memory" if the limbic system in the brain, which processes emotions and forms of memory, had been stimulated at the time of the flashback.

Nonetheless, Sigmund Freud, at the end of the nineteenth century, had noticed the unreliability of memory and had argued against localization of function. He suggested that memories are in essence fragmentary and, similar to Penfield, are only recognized as memories when they become linked to emotions. In the Freudian view, emotions structure recollections and perceptions. Our thoughts and actions are to a great extent determined by ideas, memories, and drives that are unconscious and inaccessible to conscious thought. Powerful forces (repression) prevent us from "knowing," or becoming aware of, these unconscious thoughts, memories, and drives. Yet Freud took for granted that our memories and recollections of our experiences, no matter how inaccessible to our consciousness, were carefully stored somewhere in our brains. Freudian psychoanalysis seeks to uncover these hidden thoughts, drives, and memories.

Are memories faithful to the truth, and are they accurate? If not, what are they? And if perception is based on memory, what is the nature of perception if memories are not permanent? Two very different kinds of answers have been suggested. One draws inspiration from the doctrine of localization of function as well as from computer simulations. In this view, auditory, visual, and tactile stimuli are transformed in the brain into more or less accurate representations of the physical world, and these are compared to images previously stored in the brain. The development of the computer helped

reinforce the notion of stored memories and support the belief that those memories are the sources of our motives and desires and that they are stored in different parts of the brain.

An opposing view challenges the implication that the storing of perception itself necessarily creates an accurate view of the world. Rather, it is argued that the brain, when storing perceptions, categorizes stimuli in accordance with past experiences and present needs and desires, which helps us respond to the unexpected. Our brains do not use stored images but instead use procedures that help us manipulate and understand the world—for example, procedures for recognizing the significance of letters, numbers, and objects that are encountered in a variety of different and often unrelated contexts, with different significances.

Seeing, hearing, feeling, and thinking are processes—brain processes—that cannot be understood through analogies with human mechanical inventions or the functioning of computers. Human inventions, at least as of this writing, do not have self-awareness (consciousness). No computer, no machine, be it a video camera, an automobile, an airplane, or a guided rocket, that has ever been built has self-awareness. Photographs of people are not conscious; they have no sense of who they are or who is looking at them.

No Stored, Fixed Information

There are deep reasons to reject the idea of stored, fixed information as the basis of our thoughts and actions. For if we think about our experiences, it seems unlikely that our awareness of our past and present is a re-creation of stored images in our brains—an image on a slide stored in our brains that can be projected on an internal screen. Our awareness, our consciousness, has a temporal flow, a continuity over time. Our perceptions are part of a "stream of consciousness," as the American philosopher William James suggested long ago in his book *The Principles of Psychology* (1890). Awareness (consciousness) comes from a flow of perceptions, from the relations among them (both spatial and temporal), and from the dynamic but constant relation to them of one unique perspective. Units of "knowledge" that we can transmit or record in books or images are but instant snapshots taken in a dynamic flow of uncontainable, unrepeatable, and inexpressible experience—and it is a mistake to associate these "snapshots" with "material" stored in the brain.

But what does this tell us about consciousness? Indeed, Sir John Eccles, a well-known twentieth-century Australian neurophysiologist, somewhat facetiously asked, "Why do we have to be conscious at all? We can, in principle, explain all our input-output performance in terms of the activity of neuronal circuits, and consequently consciousness seems to be absolutely unnecessary" (*Brain and Conscious Experience*, 1966, 248).

Eccles was saying that since the material processes of the brain (nerve transmission, circuits, and so on) are complete in themselves, all human behavior should be explainable in these terms. And then, of course, consciousness, an apparently nonmaterial activity of the brain, is one of those annoying ideas that one would be better off not trying to explain. It does not belong in the brain, as it is unnecessary. Chemistry and physiology can explain everything. Eccles went on to say, "I don't believe this story, of course: but at the same time, I do not know the logical answer to it" (248).

Eccles himself, in collaboration with the English philosopher Sir Karl Popper, took the position known as "dualism," which claims that there are two independent worlds—mental and physical—and that neuroanatomy and neurophysiology can only explain the physical, not the mental, world. This view has a long tradition. Its most celebrated version is that of the seventeenth-century French philosopher René Descartes, who argued that the body is material and the mind is immaterial. While it doesn't tell us much about the nature of conscious experience, the dualist view does say that it cannot be analyzed in physical terms. How the brain, a material thing, could have evolved a nonmaterial mind is not explained.

In opposition are the behaviorists, or monists, who claim that all mental experience can be analyzed in physical terms. Stimuli are physical (words are physical, simply sounds), and they cause the activation of circuits in the nervous system that lead to predetermined responses or actions. The behavioral argument was first clearly stated by the American psychologist John Watson in 1913. All behavior could be explained in terms of reflexes or learning (conditioned reflexes).

Partly in reaction to the behaviorists, a view developed that consciousness and physical processes exist in parallel. More recently, Caltech's Roger W. Sperry, the split-brain researcher, argued that consciousness is not merely a parallel activity but one that can itself influence the physical events in the brain. It is, he argues, an emergent property of the brain and can play a causal role.

Yet much of what is happening in our brains is hidden from our consciousness. We only see the end products of our brain processes. And while philosophers, psychologists, and neurophysiologists no longer believe we have pictures in our heads, the nature of our mental images remains puzzling. How is the activity of the brain related to the world around us? For the moment, at least, we have only a scant idea.

Neurological Disease

We have learned much about brain function from the effects of neurological disease and brain injury and from aberrant brain development. The effects of damage or abnormal development of a particular part of the brain give us clues as to what the brain is in fact creating. With specific forms of damage, patients can lose knowledge and understanding of the nature of colors, the nature of language, the nature of seeing. It is in this sense that brain damage can lead to the selective loss of different functions and different kinds of memory impairment. Patients with cortical blindness (their eyes function normally, but their primary cortical visual areas, which receive and process visual information, have been destroyed) lose not only the ability to see but also an understanding of what "seeing" is. Patients who have damage of the visual cortex known as V4, an area at the back of the brain that specializes in the processing of information related to color, texture, shape, and motion of objects, not only lose the ability to see colors, but they also cannot remember what colors are, or what it was like to have seen a world in color. Thus, the structure of the patient's knowledge is altered following brain damage. Patients do not lose specific memories; brain-damaged patients experience loss that is quite different from ordinary forgetting. This is shown by the story of the patient H. M.

Hippocampal Lobotomy and the Fleeting
Memories of Henry Molaison (Patient H. M.)

On September 1, 1953, William Scoville, a neurosurgeon at Hartford Hospital in Connecticut, operated on a twenty-seven-year-old man named Henry Gustav Molaison, who suffered from severe epilepsy. Scoville removed two pieces of tissue below the cerebral cortices, one on the left and one on the right side of the brain, from Molaison. These tissues, collectively called the hippocampus, are located near the center of the brain and form a part of the limbic system, which directs many bodily functions, and Scoville thought that epileptic seizures could be controlled by excising much of it. Similar operations, called lobotomies, which excised the frontal lobe, had been practiced for years, notably in 1941 on Rose Marie Kennedy, sister of the future president, who was treated for her violent rages and mood swings by Drs. James Watts and Walter Freeman of George Washington University. After the operation, Rosemary was severely disabled, unable to walk or speak, with the mental capacity of a two-year-old. H. M.'s operation differed because it removed the right and left hippocampi as well as some neighboring tissue.

H. M., as he came to be known in the medical literature (his real name was not disclosed until his death in 2008), after his surgery, could no longer remember anything he did. H. M.'s memories were fleeting or nonexistent. He could not remember what he had eaten for breakfast, lunch, or supper, nor could he find his way around the hospital. He failed to recognize hospital staff and physicians whom he had met only minutes earlier, remembering only Scoville, whom he had known since childhood. Every time he met the scientist from MIT who was studying him regularly, she had to introduce herself again. He could not even recognize himself in recent photos, thinking that the face in the image was some "old guy." Yet he was able to carry on a conversation for as long as his attention was not diverted.

H. M.'s condition suggested that the hippocampus was essential for the conversion of short-term memories to long-term memories, and he became the most widely cited example in studies of the distinction between them.

Memory Is Relational

Recent studies of how the brain organizes space and regulates how one makes sense of one's environment have shown that the hippocampus is concerned with much more than converting short-term memories into long-term memories. For example, H. M.'s sensations, thoughts, and perceptions after the operation had no continuity at all. These and other observations of scientists who studied H. M. are consistent with the hypothesis that, in the words of neuroscientists Marc W. Howard and Howard Eichenbaum of Boston University, "one of the functions of the hippocampus is to enable the learning of relationships between different stimuli experienced in the environment" ("Time and Space in the Hippocampus," *Brain Research* 1621: 345). The work of Eichenbaum and others has begun to give us not only a new view of the function of the hippocampus but a new understanding of the nature of memory. It is becoming increasingly clear that human and animal memory depend on the ability of the hippocampus to establish relations between an individual and his or her surroundings.

Studies by brain scientists including Eichenbaum and John O'Keefe, an American British neuroscientist working at University

College London, have shown that the hippocampus is made up of cells with different kinds of functions. Most important are "place" cells, discovered by O'Keefe in research that won him the Nobel Prize, which respond to an animal's location in space by causing electrical discharges or action potentials, creating mental maps of an animal's environment. These maps exist at various scales, such as maps of an entire city and maps of individual streets. "Place cells," wrote Howard and Eichenbaum in 2015, "are apparently not coding for a place *per se* but a spatial relationship relative to a landmark, or set of landmarks" ("Time and Space").

There is considerable evidence that the activities of hippocampal neurons also help establish our relationships to many other types of environmental and internal stimuli, such as sounds, odors, pain, pleasure, and fear. Howard and Eichenbaum proposed that "the spatial map in the hippocampus is a special case of a more general function in representing relationships . . . including both spatial and non-spatial [stimuli]" ("Time and Space").

In each case, the neurons are able to convey a relationship to our consciousness. The hippocampus also organizes temporal stimuli (when an event took place) and sequential stimuli (indicating the order of a series of events). The hippocampus receives and integrates many other varieties of information to create multisensory relations, which is what memory is all about.

Memories Adjust

When memories are first formed, they are "short term" and unstable. But with time, a physical representation of the memory in the

brain formed by synaptic junctions between neurons becomes more stable. This process is called consolidation. The stabilized memories then become "long-term" memories. H. M.'s brain was unable to create long-term memories. However, recent neurophysiological studies have shown that even long-term memories are very dynamic and that each time the brain tries to activate a "memory trace"— even a so-called stabilized, long-term one—the trace changes and is restabilized. In other words, memories are altered every time the brain recalls them. It is as though, were memories stored on a tape, the tape would be rerecorded in a slightly different way each time the tape was played. This alteration of an existing memory is called reconsolidation. Reconsolidation is relational as well. Because the memory trace changes, you can never remember the same thing twice in exactly the same way. You remember the memory in relation to its content upon the memory's last recall.

The process of reconsolidation, scientists have shown, changes the synaptic junctions between neurons that represent the memory. The recognition of the malleability of memory is nothing new. What is new is the observation that the connections between neurons that many scientists believe have a central part in generating memories change whenever the brain seeks to recover the information they represent. These changes may be the reason we can generalize. Over time, some memories are assimilated into categorizations or generalizations. When we recall taking the subway, we do not necessarily recall each trip separately but rather taking the subway in general, and such recollection may include an image of the subway. The brain simplifies our understanding of our environment and our relationship to it.

Hence, memory may appear to be a reproduction of images, sounds, and even thoughts that can be stored in the brain in a manner analogous to the way information can be stored on a tape, hard drive, or in the cloud, but it is becoming increasingly evident that this is too limited an understanding. Rather, as Eichenbaum, O'Keefe, and others have shown, memory is the establishment by the hippocampus (and other brain regions depending on the memory type) of complex relations among a variety of sensory stimuli from the point of view of the individual who is remembering. Thus, when Scoville removed H. M.'s hippocampus, H. M. lost more than an ability to convert short-term memories to long-term memories; he lost the ability to establish such relations.

Yet scientists still don't understand the ways that changes in synaptic junctions between neurons, or changes in cell biological or electrophysiological properties of the neurons themselves, give rise to our memories, thoughts, and actions. These mechanisms will involve the formation of what is called the engram, which is the neural substrate or physical basis for a memory—what changes in the brain when a memory is formed. Memory formation mechanisms are likely to involve changes in the structures and connections of particular neurons, and formation of even a simple memory is likely to involve vast numbers of such changes. The engram consists of, or in some other way reflects, these changes and the circuits formed by the neurons in which the changes have taken place. And as we discuss later in the chapter, we now know quite a bit about how synaptic connections change as memories are formed and can even identify which neurons are involved when particular memories are generated. However, neurobiology has yet to explain how such changes incorporate the specific information that is contained in the recall of a memory. What mechanism provides the information when we recall a particular name or emotion? Advanced techniques for imaging brain activity, such as fMRI, will reveal which brain regions are activated when a memory is recalled, but the resolution is currently far too low to study individual neurons, let alone individual synapses. Indeed, much of what we know about memory today still comes from studying the irreparable harm done to H. M.

The Physical Form of a Memory

Activation of neurons during memory formation causes changes in connections between neurons and in the neurons themselves. These changes associated with memory formation are referred to as the engram for that memory. Most important, recall of a memory is coupled with the reactivation of its engram neurons, reminiscent of Penfield's induced flashbacks.

These aspects of memory were established by studies of rats and mice. Rats and mice form memories by mechanisms similar to humans, making these rodents favorites for memory studies. Rats and mice form particularly strong memories of exposure to mild shocks applied to their feet (foot shocks). The shock-induced memories are "fear memories," and it is straightforward to tell when a rat recalls a fear memory because upon recall, the animal adopts a defensive posture and becomes motionless, or "freezes."

Researchers at MIT led by the Nobel Prize winner Susumu Tonegawa searched for neurons that form the engrams of fear memories. They placed mice in a cage and mildly shocked the mice. The following day, they returned the mice to the cage where the mice had received the shock. As expected, the mice "froze" even though there was no new shock, showing that the cage reminded them of the previous day's shock and induced freezing.

The researchers hypothesized that neurons activated during the foot shock are recruited into the engram for the new fear memory. If this were true, activation of these neurons should be sufficient to induce freezing. To test this prediction, they used an ingenious technique devised at Caltech by Karl Deisseroth and associates called "optogenetics" to activate those neurons associated with the shock, and then they observed the mice for freezing.

The opening of ion channels (pores in the membrane of a neuron that conduct ions such as sodium ions, Na^+) can activate neurons, and optogenetics is based on using a specialized ion channel called channel rhodopsin for this purpose. Channel rhodopsin has been engineered so that its pore is opened by light: simply shining light on a neuron expressing channel rhodopsin opens the channel, which in turn activates the neuron. Thus, in a living mouse, piping light through a fiber into the brain is sufficient to stimulate neurons that express channel rhodopsin.

Tonegawa engineered mice to express channel rhodopsin, but only in those neurons that were activated by a foot shock. With these mice, light would selectively activate the neurons associated with the shock, long after the shock itself had been administered. The MIT researchers shocked the engineered mice, thereby marking the neurons induced by the shock with channel rhodopsin. The following day, they delivered light through a light pipe to the hippocampus. Remarkably, when the light pulses were turned on, the mice froze! This showed that simulation of those neurons that had been activated by the shock (neurons that were marked by channel rhodopsin) was sufficient to induce fear memory recollection. The workers concluded that neurons activated by the shock became part of the engram of the foot shock memory. Notably, some of these engram neurons were located in the hippocampus, a primary memory-forming region of the brain.

Further experiments examined the role of the environment in fear conditioning. The workers asked if, during fear conditioning, neurons that were activated by the environment also become part of the fear memory engram. The Tonegawa group placed mice in a specific environment, cage A, and marked neurons that were activated by presence in that environment by expression of channel rhodopsin. In these mice, piping light into the brain would induce neuron activity normally associated with actually being in cage A. Next, they asked whether activation of these cage A–responsive neurons by light, when paired with a shock, would substitute for actually being in cage A during the shock. Would a fear memory, normally induced by being in cage A during a shock, form even though the mice had not been shocked in cage A? To test this, they placed the mice in cage B (a new location) and activated the cage A neurons with light while at the same time shocking the mice. The following day, when they placed these mice into cage A, the mice froze, even though the mice had never experienced a shock in cage A. This indicated that during foot shock training, neurons activated by the cue (by the cage A environment) become part of the fear memory engram.

Fear Memory and Hebb's Postulate

How could neurons activated by the cue become linked to the engram? The answer likely lies in a postulate made in 1949 by the Canadian neurophysiologist Donald Hebb called Hebb's Postulate. Hebb proposed that synaptic connections between two neurons that are active at the same time (connections between neurons with "coincident activity") become stronger. Considering Tonegawa's work, it appears that during foot shock training, neurons activated by the cue (the cage) and neurons activated by the shock are active at the same time. Thus, synapses linking them become stronger. As a result, neurons representing the cue (cage A)—through their linkage to the "shock" neurons—can represent the shock. They acquire the ability to induce freezing, a power normally reserved to the shock itself. This further suggests that the relational nature of fear memory (in this case, the relation of the cue to the shock) arises from the structure of the engram, in which cue-related neurons functionally substitute for shock-related neurons.

In his book *On Intelligence*, the engineer-turned-neuroscientist Jeff Hawkins observed that memories may be rich in detail: they may contain memories of place, time, and social aspects such as who was present at the time of a remembered event as well as their emotional relationships. Hawkins observed that experiencing just one of these components, such as hearing the voice of a person who was present or returning to the location of the event, which are elements of the original experience, can elicit the full memory. This evocation likely reflects the relational nature of the engram. It suggests that neurons representing each of the memory components are linked in the engram. The relational nature helps us recall valuable information acquired through our experiences, such as how a particular person behaved in a certain situation, which suggests how they might behave in a similar situation in the future. The relational nature of memory can help us decide, based on past experiences, which personal decisions are likely to lead to a good outcome or if a particular food will be tasty.

What remains elusive about the engram is exactly how a sparse group of engram neurons can represent a detailed image, or an emotion, or the text of a poem, all of which can reside in our memory. Intrinsic to this representation is how a physical group of neurons can project information into our consciousness.

Representation of Images by Brain Activity

What sorts of brain activity represent our visual cues? To start to answer this question, Dr. Jack Gallant and associates at Berkeley used functional magnetic resonance imaging (fMRI) to reveal the activities of neurons in visual areas of the cortex while a subject viewed hundreds of different natural images (images of animals, buildings, food, humans, indoor scenes, human-made objects, outdoor scenes, and textures). Next, these workers used methods of machine learning to generate an "activity pattern library," a library of brain activity patterns that were evoked by viewing particular images. When the subject had viewed a group of more than one hundred new images but focused on just one, the particular image in mind could be identified by matching the fMRI signal in the visual cortex with fMRI patterns in the library. This impressive result demonstrated that viewing different images induces different activity patterns in the visual cortex. The patterns are sufficiently distinct that they can identify which image in a group is currently being viewed. However, visual decoders can't "read your mind," even if provided with your brain's activity patterns (at least not yet). Nonetheless, this work is a step toward understanding how visual information is converted into specific neuron activity patterns.

Mnemonic Techniques

But in what sense are relationships involved in remembering the sorts of information that apparently have nothing to do with specific events or our environment, such as random lists of words and numbers? Consider, for example, the description by Alexander Luria, a Soviet Russian neuropsychologist, in his book *The Mind of a Mnemonist* (1968) of a patient, S., who could recall tables of numbers written on a blackboard. S. would study the material on the board, close his eyes, open them again for a moment, and reproduce one series from the board.

How is this ability to recall random words and numbers, even years later, related to what scientists have recently suggested is the function of the hippocampus, which is apparently essential to our capacity to remember? Luria describes how the mnemonist remembers. He never recalls arbitrary lists of words or numbers without first establishing a setting—a relation—in which he heard the lists.

> "Experiments indicated that [the mnemonist] had no difficulty reproducing any lengthy series of words whatever,

even though these had originally been presented to him a week, a month, or a year, or even many years earlier. . . ." During these test sessions S. would sit with his eyes closed, pause, then comment: ". . . You were sitting at the table and I in the rocking chair. . . . You were wearing a gray suit and you looked at me like this. . . . Now, then, I can see you saying. . . ." (trans. Lynn Solotaroff, 11–12)

In other words, the mnemonist accesses (recalls) what appear to be imprinted words only by recalling the setting in which they were first "imprinted" in his memory. Once he recalls that setting, S. has a technique that allows him to memorize arbitrary lists of numbers, words, or both. The mnemonist, Luria notes, when given a long series of words to memorize, would find some way of distributing these images of his in a mental row or sequence. Most often (and this habit persisted throughout his life), he would "distribute" them along some roadway or street he visualized in his mind. Sometimes this was a street in his hometown, which would also include the yard attached to the house he had lived in as a child and which he recalled vividly. On the other hand, he might also select a street in Moscow. Frequently he would take a mental walk along that street and slowly make his way "distributing" his images (evoked by the words) at houses, gates, and store windows. Hence, all recollections depend on a setting that the individual may or may not be aware of.

This mnemonic technique has been known since the ancient Greeks. Cicero tells us that an aristocrat named Scopas was giving a banquet, at which the poet Simonides chanted a poem in honor of his host that included "a passage in praise of Castor and Pollux."

Subsequently a note was brought to Simonides that two young men were waiting for him outside, but when he went to greet them, he did not find them. Meanwhile, the banquet hall collapsed during his absence, killing all of the guests. The corpses were badly mangled and could not be identified. Simonides remembered the place where each of the guests was sitting and was therefore able to identify them.

Simonides is generally known as the inventor of the art of memory. Most remarkable is that the art he invented operates not unlike the way in which the hippocampus creates human and animal memories by means of cells that map location in space, or create temporal markers, or encode sequences of events.

Our Recollections Adjust:
Fixed Memories Would Be Useless

A revealing way in which our memories are relational is shown by their updating, as described by the English theologian John Hull in his book *Touching the Rock* (1990). Hull became increasingly blind between the ages of twenty and his forties. When he lost his sight, he noted, "The proportion of people with no faces increased. . . . I have fairly clear pictures of many people whom I have not met again during these three years, but the pictures of the people I meet every day are becoming blurred. Why should this be?" Hull answers his own question:

> In the case of people I meet every day my relationship
> has continued beyond loss of sight, so my thoughts about

these people are full of the latest developments in our relationships. These have partly converted the portrait, which has thus become less important. In the case of somebody I know quite well but have not seen for several years, nothing has happened to take the place of the portrait, and when I think of those people, it is the portrait which comes to mind. (18–19)

Hull was losing the visual portraits of his wife and children. Hull's memories (as is true of all of our memories) were continuously being "updated." He could still visualize people he had known before he became blind and had not been in contact with since. But now that he was living in a world without any new images, his memories of people with whom he was regularly in touch were being updated into a nonvisual form—the sounds of their voices and the sensations of touching their hands and faces. When one becomes blind, the continuity of visual memory is lost.

The trauma of an accident that causes blindness can "fix" or stabilize visual imagery. The brain can no longer reorganize itself in the face of ongoing visual experience. Visual memories then seem out of joint. They are no longer tied to daily experience; they cannot be transformed by newer experiences into a nonvisual form. Visual images of people with whom the person has no contact after becoming blind remain "visual"; no new organizational patterns cause them to fade out. But images of those with whom the blind person remains in contact become transformed by new visual experiences and into a new form. Hull, who became fully blind when he was in his late forties, concluded that our memories, our recollections adjust to a changing and unpredictable world (blindness, for example). If memories were fixed, they would be useless. Recognition would be virtually impossible. You never see exactly the same person twice.

False Memories

The relational properties of memory are shown by the work of Elizabeth Loftus, professor at the University of California–Irvine, who has studied the power of suggestion and the ability to implant false events in someone's memories. In one project she persuaded people that they had undergone a traumatic childhood event in which they were lost in a shopping mall, separated from their parents. Eventually, according to this implanted memory, an elderly person rescued them, and they were reunited with their families. She implanted this memory by telling subjects that Loftus wanted to discuss events that happened in the subjects' childhood and that Loftus had prepared for the interview by speaking with the subjects' parents. Initially, Loftus and coworkers talked about things that had actually happened to establish a relation. Then the subjects were told the fabricated story. This implantation of the untrue story was repeated at three successive meetings. One-fourth of the subjects developed the false memory and accepted that it was true. In another experiment patterned on Loftus's work, investigators were able to implant the memory that the subject, when a child, went to a wedding and accidentally tipped over a punch bowl, spilling punch all over the parents of the bride. In another study, Loftus showed that the wording of questions about the collision of two automobiles altered the subject's memory of that event.

History can also be rewritten. On an August afternoon in 1972, John Stanley Wojtowicz, the son of Polish immigrants and a gay Vietnam War veteran, held up at gunpoint a branch of Chase Manhattan Bank at the intersection of Avenue P and Third Street in Gravesend, Brooklyn, New York. His objective was to obtain three thousand dollars to pay for a sex change operation for Ernest Aron, also known as Elizabeth Debbie Eden, whom he called his wife. While his accomplice, eighteen-year-old Salvatore Naturale, held female bank tellers hostage at gunpoint within the bank, Mr. Wojtowicz addressed a cheering crowd outside and spoke to a live broadcast audience. The drama, which lasted for almost fourteen hours, was observed by thousands of onlookers, and when Mr. Wojtowicz tried to make an escape in a limousine to a waiting plane at John F. Kennedy Airport, an FBI agent shot Naturale. Wojtowicz surrendered, was arrested, and served seven years of a twenty-year sentence in prison. The robbery received widespread attention and Mr. Wojtowicz corresponded frequently from his jail cell about the event.

In 1975, Sydney Lumet directed the highly popular film *Dog Day Afternoon*, which retold this story. Al Pacino played Mr. Wojtowicz and the film captured the pathos of the events, in which Mr. Wojtowicz is portrayed as an antiestablishment hero who shows concern for his hostages and attracts the support of the crowd outside.

In 1999, the French artist Pierre Huyghe exhibited in Paris a video entitled *The Third Memory* in which Mr. Wojtowicz, now released from prison, retells the story of the robbery. The video shows how Mr. Wojtowicz's recollections of the robbery, which took place twenty-seven years earlier, were radically transformed by viewing Lumet's film representation of that event. Mr. Huyghe shot his video in a television studio in Paris on a set that reconstructed the original bank. Mr. Huyghe supplied Mr. Wojtowicz with actors, and Mr. Wojtowicz assigned the roles and actions according to his recollections. "You are the bank manager," said Mr. Wojtowicz. "You go there." The video shows how the original events, which constitute a "first memory" of the robbery, and the film's re-creation of that crime, which constitute the "second memory," are synthesized to give rise to the "third memory," which was retold by Mr. Wojtowicz. As Pierre Huyghe commented, Mr. Wojtowicz "was unable to tell the difference between the relevant, which happened, and the fictional, which was a film.... He is mixing the fiction with the fact.... By telling the story, you appropriate this story, even if the story is fictional."

Thus, memory, rather than being a fixed recall of an event, is an integration of the past and present, just as John Hull integrated past and present memories of friends and relatives, and just as the moving motion picture image that we see on a screen is the consequence of the brain's integration of a series of twenty-four still photographs over a period of a second. The motion we see on the screen is memory—and consciousness.

Categorization and the Continuity of Experience

Memory consists of generalizations or categories of information and adjusts and is updated as we have new experiences, thereby creating a continuity in time and space. Artists and writers recognize the relational aspect of memory in their work, and false memories can be created by establishing false relations. We are probably much better at recognition than we are at recollection. We recognize people despite changes wrought by aging, and we recognize photographs of places we have visited and personal items we have misplaced. We can recognize paintings by Picasso and adept imitations of Picasso. When we recognize a painting that we have never seen as by Picasso or as an imitation, we are doing something more than recalling earlier impressions. We are carrying out a process recognized by the American biologist and Nobelist Gerald Edelman: perceptual categorization. As described by Edelman, categorization is a process established by the strengthening or weakening of connections of neural maps, or neural representations of our perceptions, to generate perceptual categories. In the current example, we categorize: Picassos and fakes. Our recognition of paintings or of people is the recognition of a category, not a specific item. People are never exactly what they were moments before, and objects are never seen in exactly the same way.

One possible explanation for this is that our capacity to remember is not for specific recall of an image stored somewhere in our brain; rather, it is an ability to organize the world around us into categories, some general, some specific. When we speak of a stored mental image of a friend, which image or images are we referring to? The friend doing what, when, and where? One reason why the search for memory molecules and specific information storage zones in the brain has so far been frustrating may be that they are just not there, or they may be present in unexpected forms. Unless we can understand how we categorize people and things and how we generalize, we may never understand how we remember. Yet we do remember names, telephone numbers, and words and their definitions. Are these not examples of items that must be stored in some kind of memory? Notice, however, that we generally recall names and telephone numbers in a particular context; each of our recollections is different, just as we use the same word in different sentences. These are categorical, not just specific recollections.

As we have seen from the work of Tonegawa and others, memory is likely created by linking together, at synapses, neurons that were active at the time of an event and that represent different memory components. Linkage forms a memory trace or engram in the brain, and the memory may be recalled if neurons in the engram are stimulated. Such linkage of event-related neurons may account for the relational nature of memory. Categorization may also reflect the physical form of a memory. Linkage of different engrams with similar components could give rise to categorization—for example, were memories of books or memories of subway rides or of other categories linked to one another at the level of the engram.

CHAPTER 4

· · · · ·

Altered Awareness, Self-Identity,
and the Body Image

There is more to remembering and recognizing than recalling specific events or facts. Our conscious thinking, recalling, imagining has continuity over time. Our awareness, our sense of consciousness comes from a flow of perceptions, from the relations among them (both in space and in time), and from our dynamic but constant relation to them as determined by one unique personal perspective (in other words, our subjectivity). If all human memory is relational, we might well say "The person I met today reminds me of a friend from long ago," with the relationship of today's person to the old-time friend evoking the memory.

Not surprisingly, breakdown of the ability to make relations leads to neurological disorders—and many sorts of breakdown are possible. Our ability to recognize relationships between our self and others may break down. We may also lose the ability to relate letters within words to one another, or numerals within numbers, or features of the face to form a recognizable countenance, or even the limbs of our own bodies to self. Such breakdowns give insight into the relational nature of memory.

Personal Identity

As our ideas about ourselves and our moods change, so do our feelings about who we are and what we can and will do. The search for an "identity"—made up of our constant and subtle shifts from one aspect of our personality to another in our encounters with each other—is characteristic of all of us. Neurological breakdowns can lead to syndromes of "multiple personalities"—cases in which the change from one personality to another is radical, with a loss of memory about one's previous personality. The constant changes of the body schema—a dynamic abstraction created by the brain—are at the heart of our shifting sense of identity.

What happens to an individual's self-perception when there is a breakdown of relational brain function? Such a breakdown was described by Robert Louis Stevenson in the *Strange Case of Dr. Jekyll and Mr. Hyde*: "There was something strange in my sensations, something indescribably new, and, from its novelty, incredibly sweet. I felt younger, lighter, happier in body; within which I was conscious of a heady recklessness, a current of disordered sensual images running like a millrace in my fancy. . . . I knew myself, at the first breath of this new life, to be more wicked, tenfold more wicked . . ." (111–12). And so Stevenson describes the first transformation of Dr. Jekyll into Mr. Hyde. Jekyll has become shorter, thinner, and younger. A flood of muddled sensory images overwhelms the spirit of his new body: he recognizes himself, but he is not the same person. Along with his disconcerting bodily changes, Jekyll is unable to make sense of his surroundings: "It was in vain that I looked about me, . . . something still kept insisting that I was not where I was, that I had not wakened where I seemed to be, but in the little room in Soho where I was

accustomed to sleep in the body of Edward Hyde . . ." (120). Asking himself about the source of his illusion, he concludes that while unaware, he has become Edward Hyde.

If Stevenson puts his finger on the internal fight within each of us of two or more opposed personalities, he suggests that these hostile personalities inhabit very different bodies, have very different sensations, and see and hear very different objects, places, and people. Jekyll and Hyde's pathology is a disease of consciousness—of awareness, of subjectivity, of memory, and of a sense of time and space.

And yet Hyde's pathology is not his being a murderer one moment and a researcher the next. In a curious way we all have odd fantasies that, fortunately, we rarely (or perhaps not so rarely) act upon. What is pathological about Hyde is not the evil fantasy but his having too few fantasies. Normally we slip from one personality to another without difficulty—and if we didn't, we would find ourselves hampered in our social relations. But why? Why this charade of personalities when one should do?

Multiple Personality Disorder

We all change, in apparently minor ways, from being one person at one moment to another, very different person moments later. Are we really the same person we were five weeks ago, let alone five years ago? Our brains—and our human psychologies—camouflage our shifting personalities, making us believe we are the unique person named on our driver's license.

The alternative personalities of Jekyll and Hyde represent an extreme form of personality change, a psychological condition known

as multiple personality disorder, or dissociative identity disorder. In these disorders, the individual experiences distinct identities or personality states, each with a particular sense of self and relation to the world. The person's usual identity is called the "core" identity, and the alternative personalities are called "alters." The alters may differ in many respects from the core, including in gender, behavior, interests, relationship to the world, and even ethnicity. Individuals with multiple personality disorder often have a history of abuse as a child, and their transitions are frequently accompanied by gaps in both distant and recent memories of self and may create great personal distress. The shifts in personality may, in fact, represent a means of escaping traumatic incidents of the past.

In 2018, Muhammad Rehan and colleagues, writing in the journal *Cureus*, reported the remarkable case of a fifty-five-year-old woman whose core personality could fragment into as many as seven different alter personalities. She could assume the personality of a seven-year-old child, or of a teenager, or of a person of the opposite sex (male), although most often she manifested the personality of a middle-aged woman, the personality with which she was most comfortable. When she transitioned to a seven-year-old, she became moody and arrogant and, if her wants were not fulfilled, had weeping spells or attempted to harm herself. Transition to a teenager affected her dress and her speech, which became pressured and repetitive. When she transitioned to the opposite gender, her voice and behavior changed. She dressed as a man and perceived that she had male body parts. Stress and use of drugs triggered the personality transitions, which were involuntary and afterward left the woman with no recollection of them. Rehan and colleagues proposed that the fragmentation of personality reflected "a loss of the ability to assimilate various aspects of one's identity, memory, as well as consciousness into a single multidimensional self"—in other words, a breakdown in relations ("A Strange Case of Dissociative Identity Disorder").

We believe we know who we are, and we know who we were yesterday, the day before yesterday, and a year ago. But do we? Are you really the same person you were a year ago? How often have you said aloud, or to yourself, "I've changed. I'm not the same person I was when we met. I can't explain it, but I assure you it's true," or, "I'm not myself today," or "That's not me, I don't know what's wrong"?

If you think about it, the answer is not obvious. You believe you are, of course, a particular person, let's say Joe or Jane Atkin, and we generally stop there (unless we are talking to an analyst or even a friend)—if we bother to ask ourselves who we are. So, who are you? Who is the real you? Is there a real you?

Capgras: Failure to Integrate Emotion

In 1923, Joseph Capgras and Jean Reboul-Lachaux, French psychiatrists working at the Hôpital Sainte-Anne in Paris, described a patient, Madame M, who thought her husband and children were impostors. They were, she claimed, very clever lookalikes of the members of her family. She "recognized" them, but she was not fooled by them. She "knew" her real husband and children had been kidnapped and the impostors were trying to take her money from her. Capgras and his coauthor explained Madame M's failure to recognize members of her family as a loss of an emotional reaction—

a disconnection of her emotions from her perceptions. Normally, when we see someone we know intimately, we have a conscious or unconscious emotional reaction, an emotional high, upon seeing that person. Brain damage prevented Madame M from feeling any emotions in the presence of intimate family members. Yet we believe this explanation is incomplete. Madame M had a neurological breakdown that prevented her from understanding the relations between space and time. Madame M explained, "Ça se voit dans les détails" ("It shows in the details"). Her "husband's" mustache was longer than it had been the day before, his hair was combed differently, his skin had become pale, and he was wearing a different suit.

In other words, Madame M was confronted by a parade of husbands and children whose appearances were constantly changing, however slightly. It was as if she had been looking at a series of photographs of different men and children who were impostors, pretending to be members of her family. Her brain was unable to create a visual synthesis of the constantly changing images she was seeing. It was failing to connect the images, failing to integrate them into a flow of interrelated visualizations of her husband and children that she could associate with her emotions. It appeared that Madame M had a neurological inability to integrate her family relationships over space and time.

In his book *Phantoms in the Brain*, the neuroscientist V. S. Ramachandran reports the case of Arthur, a young man who developed Capgras syndrome after suffering a brain injury in a car crash. Like Capgras's patient Madame M, Arthur insisted his parents were impostors, strangers who were simply taking care of him at home.

Ramachandran noted that when neurotypical people see their parents, the brain's face recognition system activates the autonomic nervous system, which increases heart rate and respiration and arouses the emotions, causing the palms to sweat (also called the galvanic skin response, or GSR). Using electrodes to test for palm sweating, Ramachandran verified that viewing photographs of his parents failed to activate the GSR in Arthur, a response readily observed in neurotypical individuals, indicating the lack of an emotional response. Ramachandran hypothesized that to rationalize the lack of an emotional response, Arthur created the delusion that his parents were impostors, similar to Capgras's conclusion that Madame M believed her family to be impostors because of her emotional disconnection from them. We suggest that an inability to integrate perceptions of family members over space and time accounts for the lack of an emotional response in Capgras syndrome.

Engel and Dejerine: Failure to Integrate Text and Numbers in Space and Time

Recently, the Canadian novelist Howard Engel described his own experience in the breakdown of space and time. He recalled waking up on July 31, 2001:

> What happened was I got up, went out to get the paper, and when I brought the paper in, and put the coffee on and did all the usual things, when I opened the paper, it appeared to have been written in Cyrillic or Serbo-Croatian or in an alphabet that was totally mixed between Roman and Greek characters. It was totally confusing to

me. Immediately I thought it was a practical joke. Was it April 1? I opened the paper to look inside and found that the joke had been carried to such an extent that they had faked the whole of the *Globe and Mail* of Toronto. Then when I looked at other written documents, the calendar, book covers, I found that the joke had gone too far, that something very basic was wrong. And the surprising fact was not just that the letters had gone crazy and decided to do a rhapsody of their own for their own entertainment, but that everything else was in its proper place, everything else looked as it usually did. And that there appeared to be no logic, that the hard line between reading letters and reading the view out the window was totally at odds with one another. One existed in its realm and ordinary vision existed in its. ("What Does the Brain Do? Questioning perception, Consciousness, and Free Will," https://www .dailymotion.com/video/xiqogf)

Engel continued,

When I opened a book or paper as I described already it appeared to be in Cyrillic. But the letters were also quixotic. They seemed to merge and to jump into one another. So that an Λ on first glance became a P on the second glance. And all of the alphabets seem to be changing places with one another. And there was no rhyme or reason to it. An I could look like an M and a B and a D changing places or a P and a Q. It was more thoroughgoing than that. ("What Does the Brain Do?")

This disability, called alexia sine agraphia ("alexia without agraphia"; the inability to read without loss of ability to write) arose when Engel unknowingly experienced a stroke. Indeed, alexia without agraphia is recognized to result most often from strokes that block the transfer of visual information to the language area of the brain.

A remarkably similar case was reported at the end of the nineteenth century by the French neurologist Joseph Jules Dejerine. This study, entitled "Contributions to the Study of the Anatomical-Pathological and Clinical Varieties of Verbal Blindness," continues to fascinate neurologists and philosophers to this day.

"In November 1887," Dejerine wrote, "my friend Dr. Landolt sent me at Bicêtre a patient in whom he had diagnosed verbal blindness [an inability to read] with faded and colorless vision on the right side. [The patient, Monsieur Oscar C, is] sixty-eight years old and has always been in excellent health. . . . Having been for a long time in textiles, he has acquired a small fortune. . . . His wife, who is younger than her husband by a few years, is also very cultivated. She is a particularly gifted musician and she has instilled her tastes in her husband. He frequently plays music with his wife, reading difficult scores and singing, either alone or with her. He is equally well informed about literature and reads much. . . . Monsieur C has always had excellent vision." He had created a variety of textile designs that he had drawn on graph paper. He could even count the individual threads in a fabric.

Madame C related to Dejerine that one day in late October 1887, her husband had several attacks of numbness in his right leg. The attacks continued over the following days, though her husband was able to take long walks. There was nothing really worrisome about

the attacks until the day her husband suddenly realized he couldn't read: not a single word. And yet he was perfectly able to speak and write and had no difficulty recognizing objects or people. Thinking he needed reading glasses, Monsieur Oscar C consulted an ophthalmologist fifteen days later, one Dr. Landolt, a friend of Dejerine's.

Landolt found himself confronted by a strange phenomenon that he was unable to grasp. When he asked Oscar to read the letters on the eye chart, Oscar said he couldn't read—or name—a single letter, and yet he said he could see every letter on the chart. "He claims to see them perfectly," Landolt wrote. And Oscar demonstrated the truth of his claim by copying the letters on the chart on a piece of paper. His technique of copying, however, was rather odd: "When asked to write on a paper what he sees, he recopies the letters with great difficulty, line by line, as if he were making a technical drawing, carefully examining each stroke in order to reassure himself that his drawing is exact." Nonetheless, he is incapable of naming the letters: "He compares the A to an easel, the Z to a serpent, and the P to a buckle." Oscar is perfectly aware that he is looking at letters of the alphabet. "His incapacity to express himself frightens him. He thinks that he has gone 'mad.'" He cannot read or make sense of the letters he has copied.

And yet he can write, even if, moments later, looking at what he has just written, he is unable to read his own writing: "*Now he writes from memory whatever he wants, but whether it be his own spontaneous writing or from dictation, he can never reread what he has written. Even isolated letters do not make sense to him*" (Rosenfield, *Invention of Memory* [1988], 35, trans. Rosenfield; italics in original). "Therefore, it is the sense of the muscular movement that gives rise

to the letter name. In fact, he can easily recognize letters and give their names with his eyes closed, by moving his hand in the air and following the outlines of the letters."

Oscar C's inability to read letters is paralleled by difficulties in making sense of multi-digit numbers:

> "He is able to do simple addition," Dejerine noted, "since he recognizes with relative ease single-digit numbers. However, he is very slow. He reads multi-digit numbers poorly, since he cannot recognize the value of several digits at once. When shown the number 112, he says, 'It is a 1, a 1, and a 2,' and only when he writes the multi-digit number can he say, 'one hundred and twelve.'" (35)

And though Dejerine mentions it in passing, Oscar could no longer read music, even if he could copy the individual notes, in the same way that he could "copy" the strokes of individual letters without really having an overall view of the letters, and he was unable to grasp their melodic or harmonic relations. He saw parts of the individual notes of a musical score in the same way he saw the parts of letters. He only got a sense of the relation of the parts when he closed his eyes and moved his hand in the air "following the outlines of the letters," and he probably could have done the same with music had he been asked. Finally, in what might appear to be a totally unrelated loss, in addition to his inability to read multi-digit numbers, music, letters, and words, Oscar was unable to see colors in one part of his visual field. The partial loss of color vision is, we believe, very deeply related to Oscar's problems.

After Oscar's death, Dejerine performed an autopsy on Oscar's brain, revealing brain lesions that explained, he believed, Oscar's problems: the visual centers of Oscar's brain were "disconnected," he argued, from the language centers in the brain—that is, from the centers containing the "visual images of words," an impairment that frequently results from a stroke, as with Engel's disability. Dejerine famously claimed that patients with lesions similar to Oscar's see letters and words as "drawings" that cannot be understood as language symbols because visual information never makes it to the language centers; the nerve fibers (imagine a set of electrical wires connecting the centers) are broken (disconnected). Nonetheless, Dejerine said, even though the visual centers are disconnected from the language centers, letters can still be "seen." But they are seen as drawings, not as linguistic symbols.

In an attempt to account for Oscar's ability to read single-digit numbers, Dejerine added that numbers are like "drawings," never explaining why Oscar could only read single-digit numbers and not multi-digit numbers. Single-digit numbers always represent the same number. What Dejerine's patient cannot read are multi-digit numbers, because the individual digits change their meaning (nine, ninety, thirty-nine, nine thousand) depending on their place in the multi-digit number. The same is true of musical notes; the notes can be seen but not the melodies and the harmonies. Similarly, he cannot relate letters to each other; at best he sees individual letters, but never words. In other words, the difficulties Dejerine's patient (and patients with similar brain damage) have is an inability to interconnect, to relate stimuli—to bake the cake. They are, at best, aware of relatively simple stimuli, but their brains cannot integrate them into a larger and meaningful whole.

Normally, the brain creates a visualization of whole words and multi-digit numbers. The brain is putting together individual numbers, baking a numerical cake. Without this capacity, as with Dejerine, you cannot see the relationship between different numbers; one is unable to relate the letters to one another and the notes in a score to one another.

Aphasia is a similar problem in articulating language or understanding speech. It is a problem of relating the sounds that are necessary to understand speech. It is similar to the defects in Dejerine's patient, but for sounds instead of letters.

The problems Dejerine's patient, Oscar, and Howard Engel have are that their brains are unable to create the correlations essential to perception. Just as the brain must correlate the amount of light that is reflected in at least two different wavelengths in order to create colors, the brain must correlate the lines (borders of lightness and darkness) that form letters and numbers. Oscar can read single-digit numbers, but not multi-digit numbers—just as he can make out some letters, but not words. Engel, too, cannot make sense of letters, either singly or in combinations with other letters.

Prosopagnosia:
Failure to Integrate Facial Features

The neurological inability to recognize faces, prosopagnosia, is also a consequence of the brain's inability to integrate details into larger wholes, to synthesize whole faces from "unrelated" parts of faces (noses and eyebrows). In a way similar to Engel's and Oscar's disabilities with letters and numbers, the late American artist

Chuck Close could not make visual sense of faces. He was aware of them but was unable to distinguish one from another in any meaningful way. His awareness of his surroundings, his consciousness—in particular of faces—was more limited than "normal." With prosopagnosia, when one observes a person, one is unable to relate the different images. One sees a series of static images but cannot relate them to one another. This is consistent with consciousness in normal circumstances having a "Jamesian flow," named after seminal psychologist William James's concept of a stream of consciousness that is constantly changing. As James wrote in *The Principles of Psychology* (1890), "A 'river' or a 'stream' are the metaphors by which it [consciousness] is most naturally described" (1:239). When that flow no longer exists, one cannot recognize faces or relate one moment to the next. Close's brain was unable to make sense of (and to give meaning to) the details of individual faces—much as a nearsighted person may not be able make sense of the details of a painting he or she is looking at without the aid of glasses.

Prosopagnosia is an example of an alteration of consciousness. Portrait paintings and photography, for which Close was famous, are his way of remembering faces. As a painter, Close could alter images of people's faces (greatly enlarge them using dots) and consequently alter his relationship to them. What makes Chuck Close's greatly enlarged photographs "recognizable" to him and therefore relatively easy to remember is that he could focus on details without noticing how the details are related. Close remarked, "I don't know who anyone is and have essentially no memory at all for people in real space, but when I flatten them out in a photograph, I can commit that image to memory in a way; I have almost a kind of photographic memory for flat stuff." His subjective world is one of unconnected details, and the more unconnected the details the more he could relate to them. Close, again: "Though I may be unable to recognize a particular face at a glance, I can recognize various things about a face: that there is a large nose, a pointed chin, tufted eyebrows, or protruding ears. Such features become identifying markers by which I recognize people."

Autism Spectrum Disorder:
Altered Integration of Relations

In autism spectrum disorder (ASD), too, there is a breakdown in the establishment of relations—relations with other people and with the self. Simon Baron-Cohen, Uta Frith, and Alan Leslie at the Cognitive Development Unit of University College London proposed that while neurotypical persons can readily make sense of the mental states of other persons and relate to these states, individuals with ASD lack this ability, which impairs their relations with others. We may also ask: What happens when the brain's internal images of one's own body—the brain's internal maps of the position and the varying tensions of muscles, the internal maps of heat and cold on the surface of the skin, and of touch, sound, and movement—become "mixed up," when they are no longer coordinated, no longer "in sync"? This, too, can be a problem for a child with ASD. The failure to coordinate the brain's internal mappings leads to a loss of a sense of self. The British American physicist Freeman Dyson described an artist who was on the autism spectrum whom he had known for a long time: "She has no conception of other people's existence in the way we have. It's a radically different world that she lives in. You can tell by the fact that she can't understand the difference between 'I' and 'you.' She uses the words indiscriminately" (Onnesha Roychoudhuri, "Our Rosy Future, According to Freeman Dyson," Salon.com, September 29, 2007).

The neuroscientist Michael Merzenich examined a young Indian boy, Tito Mukhopadhyay, who had no sense of his body if he wasn't moving or taking a shower. "Movement is . . . proof that I exist. I exist because I move." Merzenich found that Tito, as in other cases of children on the spectrum, have "mixed-up brain maps." It happens that, when asked to point to their nose, they may point to their ear. Children with ASD often have clumsy movements. The failure of the brain to correlate sensory information is shown by Tito's remark: "I can concentrate on what I see, or on what I hear, or what I feel . . . I never thought of this as abnormal until I realized that the others can simultaneously see, hear, and feel . . . I need time to prepare my eyes, I need time to prepare my ears. Otherwise, the world is in chaos . . ." (Sandra Blakeslee, "A Boy, a Mother and a Rare Map of Autism's World," *New York Times*, November 19, 2002).

Similarly, Temple Grandin, the Colorado State University animal behaviorist who is also on the autism spectrum, described her visualization of a type of comfort machine while daydreaming as a child:

> This design was sort of a coffin-like box. I imagined crawling in the open end. Once inside, I would lay on my back, inflate a plastic lining which would hold me tightly, but every so gently. And most importantly, even in my

imagination, I control the amount of pressure exerted by the plastic lining. (Grandin and Margaret M. Scariano, *Emergence: Labeled Autistic*, 30)

This was later realized as a "compressing machine," "hug box," or "squeeze machine" and consisted of two padded plywood boards that pressed, through the action of compressed air, against the individual to give the feeling of being held by another person and thus comfort them. The physician and author Oliver Sacks described its effect:

It is not just pleasure or relaxation that Temple gets from the machine but, she maintains, a feeling for others. As she lies in her machine, she says, her thoughts often turn to her mother, her favorite aunt, her teachers. She feels their love for her, and hers for them. She feels that the machine opens an otherwise closed emotional world and allows her, almost teaches her, to feel empathy for others. ("An Anthropologist on Mars," *New Yorker*, December 27, 1993, 106)

Curiously, persons on the spectrum may show great artistic talent, as if compensating for their difficulty to relate to other people. However, they may draw or paint only inanimate objects, rather than people, as seen in the work of the young Brazilian artist Camila F., who is on the spectrum and paints exuberant and colorful canvases: *The Rainbow* (O Arco Iris), *Squares* (Quadrados), *Circle of Polka Dots* (Ciclo de Bolinhas), and other abstract compositions. Also, persons with ASD often relate to detailed lists. They are comfortable with details but not the whole. A young man who is on the spectrum, when he reenters a dining room where the table is set, notices that a spoon has been moved. A woman sees only one shade of red. People with ASD often see detail—rather than the overview.

Creation of the Body Image

Memory and recognition depend on our subjective state, and what we mean by our subjective state is our relation to our self, to others, to our surroundings, and to our past, present, and future. The shifting personalities of Jekyll and Hyde are changes in subjectivity: a person sees himself at one time as Jekyll and at another as Hyde. Memory is relational, and crucial to the establishment of these relations is the creation by the brain of the body image. Furthermore, all of our memories are subjective—they are created from the point of view of the individual who is remembering. We have a sense of self because we have a preexisting sense of our body that contains that self. The basis of our subjectivity is our "body image," a coherent, highly dynamic (it is constantly changing with our movements), three-dimensional representation of the body in the brain. This body image is an abstraction the brain creates from our movements and from the sensory responses elicited by those movements—using one's left hand to pick up an apple, for example. "The coherence of consciousness through time and space is again related to the experience of the body by way of the body image," the American philosopher John Searle wrote in 1995. "Without memory there is no coherent consciousness" ("The Mystery of Consciousness: Part II," *New York Review of Books*, November 16, 1995).

Apparently, a normal body image is essential to the formation of our perceptions, our memories, and our awareness of our past and present. The body image serves as a frame of reference for our perceptual and mental worlds. It may be part of the innate dynamic of the brain, for our brains must "represent" our bodies—the muscles, the pain receptors, the temperature receptors, the position receptors ("proprioception"), the movements—for us to be able to function. We are moving organisms confronted by unpredictable, messy environments, and our survival requires the brain to create some kind of coherence so that our actions "make sense" in the immediate world and will probably make sense moments from now in the immediate future. It does this, in part, by trying to preserve an intact body image. Indeed, an intact body image is essential for our normal mental worlds.

How does the brain create a body image? Since the nineteenth century, it has been known that the brain creates "maps" of the body in the cortex. There is a cortical map of sensations (a sensory map) and a cortical map of movement (a motor map). In the sensory map, the region in the brain that is activated—for example, by touching the hand, fingers, and arm (the cortical area that "represents" the sensations created by a cotton swab moved from the tip of the fingers to the arm)—is adjacent to the representation of the face.

Alien and Phantom Limbs

Since our subjectivity depends on our body image, if our body image is altered for neurological reasons, so, too, are our bodily recollections. After he badly injured his leg on a mountain in Norway, Oliver

Sacks described what is known as the "alien limb" phenomenon in his book *A Leg to Stand On* (1984): "The leg had vanished, taking its 'place' with it. Thus, there seemed no possibility of recovering it.... Could memory help, where looking forward could not? No! The leg had vanished, taking its 'past' away with it! I could no longer remember having a leg. I could no longer remember how I had ever walked and climbed" (58).

What would happen if one actually lost an arm or a leg? Can a person's changing sense of his or her own body cause memories to change and disappear? Can physical damage to our bodies affect our thoughts and actions? Does the very real change in body image cause a change in the way we see and understand our surroundings or the world around us? It certainly changes the way we see ourselves. If we lose a hand, an arm, or a leg, our brain often creates a "phantom limb"; that is, it creates a "sensation" of the missing limb. For example, it will create the sensation of a hand more or less where the hand normally would be.

A phantom limb refers to the perception by an amputee who has lost an arm or leg in surgery or in an accident that the limb is still attached to the body. The phantom limb might be extremely painful. When points remote from the amputation line are touched, such as the amputee's face, he or she paradoxically feels a phantom limb. Remarkably, memories related to the original limb may be linked to the phantom limb. The subject may even perceive that the phantom limb is wearing a wedding ring or jewelry; when the weather turns humid, the phantom limb may experience arthritic pain. The patient's phantom limb is not only a recollection of the lost arm or leg but one that includes the patient's experiences related to that limb.

And, as if the brain wants to make sure the patient knows the limb does not really exist, it creates phantom pain to accompany the phantom limb. It is as if the brain wants to make sure the patient knows there is nothing where the phantom has been created. Pain is our brain's way of pointing to sensations that don't make any sense. Phantom pain is the result of the incoherence between what the brain "sees" (no arm) and what the brain "feels" (the presence of a phantom).

Why a "phantom"? Why this odd creation? In creating a phantom the brain is apparently attempting to restore a unified sense of self and continuity with the past. When a phantom limb develops (almost always following an accident), the individual's memories and his awareness of his surroundings remain "normal." The phantom is abnormal, but it is the price one pays to keep the rest of our perceptual world normal. Why?

Children born without a part of their arm or leg may, perhaps surprisingly, also have sensations in the "missing" limb. We are born with brains that have an internalized "normal" body image that does not necessarily correspond to the actual state of our bodies. And since our bodies are constantly changing—we walk, run, write, dance, play musical instruments, age, and become ill—so, too, are the ways in which our brains "represent" our bodies. From the brain's point of view, the "body image" is the highly dynamic (ever-changing) neuronal activity that apparently forms the background, the frame of reference, for its other activities, including its creation of our memories.

Indeed, the brain behaves as if it were desperate to somehow create a "normal" body image even if we have lost an arm or a leg. The neurologist V. S. Ramachandran had a patient who had lost his hand in a motorcycle accident. Ramachandran devised a therapeutic procedure that successfully eliminated phantom pain. The patient put his intact hand in one side of a box and "inserted" his phantom hand in the other side. One section of the box had a vertical mirror, which showed a reflection of his intact hand. The patient observed in the mirror the image of his real hand and was then asked to make similar movements with both "hands," which suggested to the brain real movement from the lost hand. Suddenly the pain disappeared. Though the young man was perfectly aware of the trick being played on him—the stump of his amputated arm was lying in one section of the box—the visual image that looked, due to the mirror, like intact hands on both sides overcame his sense of being tricked. Seeing is believing! Pain—which, as we have said, is the consequence of the incoherence between the brain's creation of a phantom limb and the visual realization that the limb does not exist—disappeared, and what was seen (a hand in the mirror) matched what was felt (a phantom).

At one moment, then, the patient experiences a painful phantom limb, and at another he sees a mirror image of his intact hand and the pain disappears. Much like Dr. Jekyll and Mr. Hyde, the patient in the experiment sees and remembers one world at certain times and a completely different world at other times. The phantom limb is the brain's way of preserving a normal body image—a sense of self that is essential to all coherent brain activity.

The brain feels a necessity to create a complete body image. When

it sees two body parts juxtaposed in a seemingly normal way, it incorporates them into the body image. In an experiment performed in the 1960s, subjects were asked to introduce a gloved hand into a box. They were told to observe their hand, but they were not informed that another gloved hand had been introduced into the box just above theirs. The gloved hand they were actually observing was that of the experimenter, not their own. They were then told to make certain movements ("make a fist, now open it") with their hand. The experimenter made precisely these movements and the subjects believed they were watching their own hand. From time to time, the experimenter failed to follow the commands and the subject saw his gloved hand moving in a way that was different from what he was actually doing. For example, if he had been told to make a fist, the experimenter spread his hand open. About 30 percent of the subjects believed the hand they were observing was their own and they felt they were being controlled by an external force. They also felt considerable pain when their hand "failed" to carry out the experimenter's commands. When the experiment was performed on patients with symptoms of schizophrenia, 80 percent of these patients complained their hands were being controlled by an outside force.

The source of the pain the subjects felt when they observed the experimenter's hand disobeying the commands appears to be related to the incoherence between what is being observed (a hand spreading wide open) and the action the brain believes it is performing (making a fist). When what the brain "sees" and what it "feels" it is doing appear to be identical, there is no pain, no feeling that an outside force is controlling one's movements.

Awareness of Self by Animals and Children:
The Rouge Test

The body image serves as a frame of reference for our relations to self and to the external world. The brain tries to maintain the intactness of the body image by generating a phantom limb in the case of an amputee, or by interpreting another person's body part (the gloved hand) as its own when perceptions of the body part indicate self. These behaviors suggest strong mechanisms in humans for establishing and maintaining awareness of self. We might ask: Are animals also capable of self-consciousness or self-awareness, and when during development do humans acquire this ability?

To test an animal for self-awareness, psychologists developed the mirror self-recognition test, or rouge test. In this test, an animal is anesthetized, and a dot of odorless, rapidly drying red ink is applied to the animal's forehead or other location where the animal cannot see the mark directly. After awakening, the animal is presented with a mirror and is observed for actions that suggest that the animal associates the image seen in the mirror (which has a red dot) with self. Such an association would be indicated by the animal examining its own forehead for the red dot or orienting itself toward the mirror to make the dot more visible. Elephants, dolphins, and chimpanzees, which are among the more intelligent animals, pass the test, thus indicating that they are self-aware.

Infants (children of six to twelve months) fail this test of self-awareness, while children at twenty months pass the test. This indicates that self-awareness is a higher brain function that, in humans, is established at a time of rapid brain development, when children begin to exhibit fantasy play, independence, and increased vocabulary.

Thus, we see that the brain establishes our relations to self and to our environment. We all search for self-identity and experience subtle shifts in mood and feelings, but should the brain be unable to establish relations, the shifts may be great and result in neurological dysfunctions such as multiple personalities and the inability to integrate emotions that exist between us and others, to recognize the significance of integers within numbers and letters within words, or to integrate facial features during recognition of another person. All of these failures can render one incapable of productively interacting with one's social or physical environments. Persons on the autism spectrum also manifest altered integration of social relations. In the search for self-identity, we form a body image, and if the body image is disrupted, such as by loss of a limb, the brain takes actions to restore it by creating a phantom replacement, emphasizing the importance of the body image to establishing relationships with self and the world.

The Continuity of Experience

Our perceptions are part of a continuity of experience. Our sense of color, or of smell, or of motion comes precisely from the flow of perceptions and from the comparisons the brain makes from moment to moment. For example, motion pictures give us a sense of continual movement by means of a series of static images presented in rapid succession. Our conscious experience is not one of one static image followed by another. Instead, we see motion because our brains create motion by relating one image to the next. In general, it is this relating, this connecting between moments, not the moments themselves, that is at the heart of our perceptions and recollections. Conscious perception takes place over time: the continuity of our sense of space, of time, and of our recollections derives from the correspondence the brain establishes from moment to unrelated moment. Apparently static images are not static but relations, connections the brain establishes over time. Without this activity of connecting, we would merely perceive a sequence of unrelated stimuli from moment to unrelated moment, and we would be unable to transform this experience into knowledge and understanding of the world.

More than two hundred years ago, the Scottish philosopher David Hume caught the essence of the nature of our consciousness and our recollections when he wrote, "Our thought is fluctuating, uncertain, fleeting, successive, and compounded; and were we to remove these circumstances, we absolutely annihilate its essence" ("Dialogues Concerning Natural Religion," 1908, 58).

It is the "fluctuating, uncertain, fleeting" nature of consciousness that Picasso and Braque were trying to capture in some of their early Cubist paintings. And when Marcel Proust published the first volume of his novel *À la recherche du temps perdu* (In Search of Lost Time), he, too, used "uncertain, fleeting" awareness as one of his principal themes. When driving through the countryside, Proust's narrator sees the twin steeples of Martinville in the distance. Suddenly, a third steeple, Vieuxvicq, springs into view. The narrator is unsure how to judge the distance between him and the steeples. He suspects they are a long way ahead, when the car swerves around a

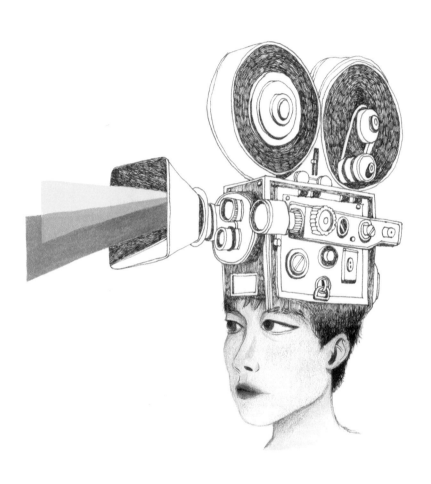

bend and almost sets the steeples down at his feet: "They had flung themselves so abruptly in our path that we had barely time to stop before being dashed against the porch of the church" (1992, 256). Our perceptions of people and things vary from moment to moment, Proust writes, "like different perspectives in a countryside where a hill or a castle seems at one moment to be to the right, at another to the left, to dominate a forest or emerge from a valley, thus reminding the traveller of changes of direction and altitude in the road he has been following" (*In Search of Lost Time [Complete Volumes]*, trans. Sydney Schiff [2016], Internet Archives, 7:2612). Just as we have no fixed view of the landscape, there is no fixed way we understand each other. Human psychology is fluctuating and uncertain. And yet while we are looking at constantly changing perspectives, we are conscious of unified images. When we view twenty-four still images of a movie flashed on the screen every second, we see motion: actors, cars, robots, and missiles, all moving. Yet there is no movement on the screen. The movement is entirely an invention of our brains.

CHAPTER 5

. . . . .

Perspectives on Future
Understanding of the Brain

The shift from the idea of fixed, stored memories that are internal re-creations of an external reality to the new view of memory as relational is just one of many changes that were made over the past century as scientific dogmas were revisited and more information was gathered. In physics, the realization by Einstein of the relational nature of time and space led to the recognition that classical Newtonian rules governing the movement of macroscopic objects were oversimplified, and to the discovery of quantum mechanics, which holds that subatomic particles behave in unpredictable ways. Likewise, our understanding of genetics has changed greatly since the time that Mendelian inheritance and the structure of DNA were thought to explain completely the gene and its functions.

The former belief that the laws governing subatomic particles are similar to those governing large-scale phenomena is not unlike the fallacy that there are fixed images stored in our brains. We have known for over a century that the Newtonian laws of motion is not valid at the subatomic level. What appear to be smooth movements of planets in orbit around the sun are, when considered at the level of electrons and other subatomic particles, apparently actually "jittery" quantal movements. By analogy, at the level of the conscious individual, one is aware of coherent objects, and yet at the neuronal level there are, as yet, no identified representations of objects or thoughts. Indeed, if geometry is the "cement" of the Einsteinian universe (as the law of gravity is the cement of the Newtonian universe), by analogy we might say consciousness is the cement of brain activity.

At the beginning of the twentieth century, artists and writers also recognized the relational nature of memory and argued that an understanding of our ever-changing psychology depends on our ability to relate space and time. What is essential to understanding our ever-changing psychology is that the ability to relate space and time is a critical part of the normal functioning of memory and consciousness. The breakdown of the brain's ability to establish a relationship between space and time alters an individual's consciousness.

In 1902, a few years before Proust's novel *À la recherche du temps perdu* appeared, the French mathematician Henri Poincaré published *La Science et l'Hypothèse*, the first book of popular science to be published in France. Proust and the Cubists were probably aware of Poincaré's book.

Before the publication of Poincaré's book, the Newtonian view of the universe, that space and time are absolute, was generally held to be true: a ruler measuring one meter would measure exactly one meter anywhere in the universe, just as anywhere in the universe a clock would run at exactly the same speed. With his book, Poincaré challenged these ideas by writing that there is no absolute space and no absolute time. In 1905, Albert Einstein published his theory of special relativity, claiming, like Poincaré, that there is no absolute space or time. Furthermore, in Einstein's paper, the speed of light is absolute for all observers. It is the constancy of the speed of light that makes space and time relative.

In 1908, the German mathematician Hermann Minkowski reformulated Poincaré and Einstein in his famous lecture "Space-Time," declaring: "The views of space and time which I wish to lay before you have sprung from the soil of experimental physics, and therein lies their strength. They are radical. Henceforth, space by itself, and time by itself, are doomed to fade away into mere shadows, and only a kind of union of the two will preserve an independent reality" (Hendrik A. Lorentz et al., *Principle of Relativity*, 1952, 75). Minkowski's unification of space and time into a four-dimensional continuum is the basis of all subsequent work in the theory of relativity and made possible Einstein's 1916 theory of general relativity.

Proust may well have been referring to Poincaré (and perhaps to Einstein and Minkowski) on November 12, 1913, two days before publication of the first volume of *À la recherche du temps perdu*, when he published an "interview" with himself describing his conception of his novel: "We have both plane and solid geometry—geometry in two-dimensional and three-dimensional space [Euclidian and non-Euclidian geometry]. Well, for me the novel means not just plane psychology [Euclidian psychology] but psychology in time [in space-time]. It is this invisible substance of time that I have tried to isolate" (Roger Shattuck, *Marcel Proust*, 1974, 169).

Parallel to the changes in our understanding of physics, there is a new understanding of genetics that has changed our conception of the gene and the relationship of DNA to genetic mechanisms. The early twentieth-century view, called the modern synthesis, was a fusion of Darwin's theory of natural selection with Mendelian theories of inheritance. The modern synthesis gained mechanistic insight with the rise of the field of molecular genetics. In 1941, George Beadle and Edward Tatum, at Stanford, observed that gene mutations in the mold *Neurospora crassa* most often affected only one step in a metabolic pathway, with each step carried out by a particular enzyme, and thus they proposed "one gene, one enzyme." This quite simple relationship—that a gene encodes a single protein (enzymes are proteins)—failed to recognize the actual great complexity of gene function, much as Newtonian physics failed to represent the complexity of quantum physics. It is now evident that a single gene (even the unitary concept of a gene has been challenged) can encode many different proteins through the mechanism of messenger RNA (mRNA) splicing, discovered by Phillip Sharp at MIT and Richard Roberts at Cold Spring Harbor. Splicing makes possible the encoding of many different mRNAs from a single stretch of DNA, with each mRNA giving rise to a different protein. Further, we now know that stretches of DNA can break and rejoin at sites within a gene, as seen for genes encoding antibodies, thereby reorganizing genes at the DNA level. We also now appreciate that not only the *way* information is stored in DNA but the *way it is utilized* is critical for how genes function. Newly understood mechanisms, called epigenetic mechanisms, can profoundly alter gene expression (which genes are transcribed into mRNA and thus translated into proteins) without modifying the DNA-encoded information itself, and in a sense provide DNA with a form of memory of its own. Thus, the different components of genetics are not independent of one another. The notion "one gene, one enzyme" obscures the relational aspect of genetics, in which the function of a gene is related to many factors, including the status of the splicing of its mRNA and the status of epigenetic control of its expression. Thus, just as our sensations and memories arise from relations, the complex functions of genomes arise from relations, and similarly, the speed of light establishes the relationship of space and time.

Perspective on the Future

In the coming years, new techniques for genetically modifying brain function and for controlling the activities of specific neurons will provide powerful means for studying the physical brain-consciousness relationship. The first indications from such studies are that engrams—memory traces in the brain—for particular memories are composed of complex networks of neurons that extend to many brain regions, a feature of the engram that may reflect the diversity of information that resides in a memory. This complexity of engram circuits may also enable the brain to establish relations and to categorize by representing in the circuitry subtle relationships that reside within sensory information. To begin to understand how the brain creates relations, we may investigate how manipulations of neuron circuits and neurons themselves alter the engram for a memory of a face or a family member, and how engrams are changed in neurological disorders, such as prosopagnosia or Capgras syndrome.

Functional magnetic resonance imaging (fMRI) will complement the new genetic techniques by revealing the actual brain activity that correlates with a particular conscious state. While fMRI has only limited precision, we may enhance this precision by using "deep-learning" statistical computational techniques to extract high-resolution activity patterns related to particular brain states from big sets of fMRI data. This will reveal new aspects of the physical brain-consciousness relationship. So far, such studies have been limited to correlations: to matching an experimental subject's fMRI pattern to fMRI patterns recorded from an individual whose brain consciousness content was known. However, at the time of this writing, we cannot read a person's thoughts by inspecting an fMRI image of their brain. We can only say that an fMRI image pattern looks like one recorded from a person who we know had a particular thought in mind. But even these limited successes will encourage us to uncover the rules or algorithms that relate the physical brain to the information that resides in consciousness.

The development of noninvasive brain-computer interfaces (BCIs) that allow exchange of information between a brain and a machine may help us overcome a major barrier to understanding consciousness, which is that our conscious thoughts are accessible only to ourselves. When coupled with high-resolution fMRI and use of artificial intelligence and machine learning procedures to interpret the brain activity data obtained with BCIs, we may begin to understand how information is encoded in physical brain structures.

A completely different approach to studying the brain uses brain organoids, self-assembled, three-dimensional aggregates of cells about a millimeter across derived from stem cells (cells capable of differentiating into many specialized cell types). Organoids grown in culture in three-dimensional bioreactors resemble miniature embryonic human brains. By exposing growing organoids to chemical agents and growth factors, we may mimic normal development of specific brain regions, such as the cortex or hippocampus. Recently, cortical organoids have been produced that display coordinated brain activity resembling that of the embryonic brain and that could possibly regulate their circuits by Hebbian mechanisms, analogous to the mechanisms of memory formation. Could organoids be used to study basic features of memory formation and consciousness? Although achieving true consciousness in a cultured tissue is unlikely, the scientific community is debating the ethical issues connected with the creation, in vitro, of some aspect of consciousness.

Psychedelics, drugs that alter the psychological, visual, and auditory states of consciousness, also offer new opportunities. Psychedelics alter consciousness by releasing the constraints on what we accept as reality, leading to hallucinations, and psychedelic drugs are now being employed in clinical psychotherapy as treatment for depression and PTSD. Psychedelics alter sensory perception and the perception of self and of reality, brain processes proposed here to be used by the brain to simplify our sensory world. Thus, psychedelics, coupled with brain imaging and psychological assessment, may serve as laboratory probes for understanding how the brain generates conscious perceptions.

It is likely that many new and not yet anticipated approaches for studying consciousness will emerge, given the great energy of the

research scientist community. However, it is still too early to anticipate the impact of such future activity on resolving the most challenging questions, such as the "hard problem of consciousness" or the physical nature of our subjective experiences. Certainly an explanation is not assured, but we may note that other hard problems, such the wave–particle dual nature of matter and the physical basis of heredity, have been resolved, at least in large part.

The Brain Takes Us from Chaos to Stability

So, what has been misunderstood about the brain functions that enable us to survive in a chaotic and unlabeled world? A deep clue comes from the realization of the significance that plants don't have brains. Only animals—and even very primitive animals and insects—have brains. Brains evolved because moving creatures, no matter how simple, are confronted by ever-changing, unpredictable surroundings. Plants don't have brains because they don't need them; they don't move from place to place. For animals, motion creates a world of visual, tactile, and auditory sensations that are unorganized and unstable; in short, for animals the world is constantly changing. What the brain must do—it's probably the principal reason brains evolved—is create a stable, coherent sensory environment for the individual organism to understand and use. While our proposal represents a step in the evolution of thinking about brain function, this thinking is far from complete. There is much to be learned and understood about how the brain invents our perceptions, of what we see, hear, and feel when we look, listen, and touch, as we explore our sensory environment.

APPENDIX

A Basic Overview of the Brain

To assist the reader new to the field of neuroscience, we have provided a basic overview starting with the brain's anatomical organization and how it serves the processing of sensory information. We go on to discuss the brain's primary constituent cells—the neurons—and how they transmit signals within the brain.

Neuroanatomy and Brain Function: Evolving Conceptions

Throughout history, physicians and philosophers have speculated about the brain. To the early Greeks, the brain appeared to be a collection of pumps joined by tubes, and the seventeenth-century philosopher and mathematician René Descartes suggested that ventricles, fluid-filled cavities in the brain, pumped fluid via the nerves to inflate the muscles, generating movement.

Modern interpretations of brain anatomy and function have evolved greatly. Anatomically, the brain is dominated by its most prominent feature, the cerebral cortex, which in humans makes up more than 70 percent of the total brain substance and is a site of higher processing of sensory information and formation of motor activity. Because cortical regions become active as one becomes aware of features of one's environment, the cortex is considered to be a major seat of consciousness.

The cerebral cortex consists of two cerebral hemispheres, the right hemisphere and the left hemisphere, which are highly convoluted structures folded into sheets that pack the large surface area of the cortex into a small space. Physically, the two hemispheres are separated from each other by a deep crevice, and each hemisphere is bisected by a second deep groove called the central sulcus. Posterior to the central sulcis lie three anatomical regions called lobes: the parietal, occipital, and temporal lobes, which make up the somatosensory cortex and the association cortices, structures that process and integrate sensory information. Anterior to the central sulcis lies a fourth lobe, the frontal lobe. The anterior (front-most) portion of the frontal lobe, the frontal cortex, carries out executive functions such as reasoning, judgment, and social understanding, and the posterior portion contains the primary motor areas of the brain, which plan and execute voluntary movement. Stimulation of the motor region of the frontal lobe makes muscles contract, while stimulation of the somatosensory cortex, which faces the motor region across the central sulcus, produces a parallel pattern of tactile sensation, rather than motion.

Processing Sensory Information

We receive information about the world through interaction with its physical features: photons, vibrating waves of air, volatile small molecules, and so forth. These interactions take place through

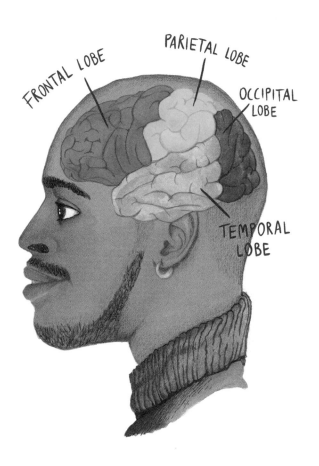

specialized proteins, called receptors, that are embedded in the surface of neurons in sensory organs (retinae, nasal cavities, etc.) and that are exposed to the environment. Physical features of the environment (photons, airwaves) activate particular types of receptors located in the relevant sensory organs (eye, ears), which creates nerve impulses that travel through a circuit of neurons to the sensory cortex for processing.

Information from each sense is processed in its own region of the cortex. Visual stimuli are processed by the visual cortex, which is located in the occipital lobe at the back of the brain. The olfactory cortex at the base of the forebrain, in the frontal lobe, processes odor information, and the auditory cortex, located in the upper region of the temporal lobe, processes sound information, and so on.

After initial processing of the sensory information in a primary region, processing continues in secondary and yet higher order cortical processing regions, in which finer and finer details of the sensory information are extracted. Ultimately, the information for the different senses is integrated in the association cortex with other information, both internal and external, about emotion, pain, feelings of pleasure, and so on to form a multisensory representation of the environment.

The prefrontal cortex, which is perhaps the most highly evolved region of the human brain, lies in the frontal lobe, adjacent to the motor cortex, and receives highly processed information from other cortical areas and other basic brain parts. The prefrontal cortex weighs incoming information contained in memories of past experiences together with sensory information from the present to help select behaviors that are most appropriate for the moment.

The limbic system, a group of structures below the cortex, also processes sensory information and is important for emotional and sexual behavior as well as formation of memory. One limbic structure, the hippocampus, is the rolled inner edge of the cerebral hemispheres (in cross section giving the appearance of a seahorse— hence its name). It extracts from brain activity information about spatial and social relations, such as our position within our physical environment or within a social hierarchy. Other limbic structures are the amygdala, which processes information related to fearful situations, and the hypothalamus, which regulates thirst, temperature, and hunger as well as sleep and emotions and other basic body functions.

Neuron Function: Synapses and Circuits

The brain is composed of two main cell types: neurons, which are specialized cells that receive and transmit information by electrochemical mechanisms, and glia, which are intermingled with the neurons and provide support for neuron survival and function.

Neurons are highly asymmetric cells that transmit signals to other neurons at sites of contact called synapses, also called synaptic junctions. A series of neurons, one connected to the next at synapses, forms a circuit. Synapses are formed by contact of a presynaptic neuron, which initiates signaling and lies upstream of the synapse, with a postsynaptic neuron that receives synaptic signals and lies downstream of the synapse. The synaptic signals are transmitted in the form of neurotransmitter molecules, which are chemicals that are released from the presynaptic neuron and that travel to receptors on the postsynaptic neuron. (More about neurotransmission later). For many, but not all, neurons, these receptors are located on the surface of mushroom-shaped structures called spines. Spines in turn are positioned on dendrites. There are many dendrites, each supporting synapses positioned on spines. The dendrites form branched structures that resemble tree branches, and collectively they form what is called the dendritic arbor of the neuron.

The dendritic arbor is connected to the neuron cell body, which contains the nucleus and the bulk of the neuron's cytoplasm. Extending from the cell body is the axon, a long structure whose tip, the axon terminal, forms the presynapse and releases neurotransmitter molecules that travel across the synapse and convey signals to the postsynaptic neuron, which is the next neuron in the circuit.

Neurotransmitters and Chemical Transmission at Synapses

Signaling by neurons takes two general forms: (1) signals travel within a neuron, down the length of the neuron, and (2) signals travel from one neuron to the next, across synapses. The signals that are conducted by a neuron down its length take the form of nerve impulses, electrochemical waves that are generated by an electrophysiological mechanism called the action potential. We understand the action potential through the work of numerous researchers, notably Bernard Katz, Alan Hodgkin, and Andrew Huxley, who worked in England from the 1930s to the 1950s. Nerve impulses are initiated when neurotransmitter molecules are released

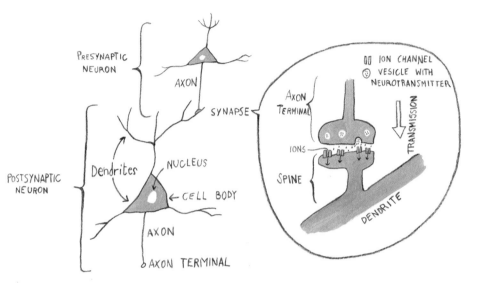

PRESYNAPTIC NEURON

AXON

SYNAPSE

POSTSYNAPTIC NEURON

Dendrites

NUCLEUS

← CELL BODY

AXON

AXON TERMINAL

ION CHANNEL
VESICLE WITH NEUROTRANSMITTER

AXON TERMINAL

IONS

SPINE

TRANSMISSION

DENDRITE

from the presynaptic neuron. Release is from a blob-like structure called the axon terminal or bouton (French for "button"). Under the electron microscope, small vesicles can be seen in the bouton, and inside each of these vesicles are about one thousand molecules of transmitter. The bouton is separated from the downstream, post-synaptic neuron by a gap of twenty nanometers called the synaptic cleft. When a nerve impulse arrives at the bouton, one or more vesicles attaches to the presynaptic membrane and then spills its contents into the synaptic cleft. As shown by Katz in 1952, the release is "quantal"—about the same number of neurotransmitter molecules are released from each vesicle.

After release, the neurotransmitter molecules diffuse across the synaptic cleft. When they arrive at the membrane of the postsynaptic neuron, they bind to neurotransmitter receptors, proteins that form pores in the synaptic membrane. Upon binding a neurotransmitter, the receptors undergo a conformational change that opens their pores. Ions flow through the open pores, into and out of the neuron, with the overall direction of the flow depending on the ion's charge—positive or negative—and the ion's relative concentrations inside and outside the cell. Because ions carry an electrical charge, the flow of ions changes the magnitude of the electrical charge within the neuron, which regulates the activity of the neuron (its likelihood to form new nerve impulses or action potentials).

All cells, including neurons, are bathed in a fluid that bears a close resemblance to seawater. (This may be related to the probable origin of life in the sea.) The external fluid contains a high concentration of positively charged sodium ions and negatively charged chloride ions (when combined, sodium and chloride ions become common table salt), whereas the fluid inside the cell contains high concentrations of positively charged potassium ions. Neurons take advantage of this ion imbalance to transmit signals rapidly down their length. The pores of different receptors in the nerve cell membrane are of different sizes. Some allow only sodium ions to cross the membrane and flow into the neuron while others permit passage of potassium ions out of the neuron. Yet others admit chloride to flow into the neuron.

The pores are a part of protein structures called ion channels. The opening of a pore is controlled by one of two mechanisms. Some pores are opened by the binding of neurotransmitters, as described earlier. In this case, the ion channel is called a "receptor" (because it binds or receives a neurotransmitter) and its activation is said to be neurotransmitter-gated or ligand-gated (gated means regulated). The second type of ion channel is opened by changes in the voltage gradient across the neuronal cell membrane. Such ion channels are said to be voltage-gated. You can measure this voltage gradient with a voltmeter by placing one electrode into the cytoplasm of the neuron and the other electrode into the fluid surrounding the neuron, where the voltage potential is defined as 0 millivolts (mV). When neurons are at rest, the voltage gradient measured in this manner is usually about −60 mV.

Neurons use the flow of ions across the neuron's membrane to change the voltage gradient across the membrane—that is, how negatively charged is the interior of the neuron. As positively charged ions (sodium ions) flow in, the neuron interior becomes less negative and is said to be depolarized, a state in which most neurons are prone to be active. As potassium flows out or chloride flows in, the interior becomes more negative, or is said to become hyperpolarized, and the likelihood of neuron activity decreases. Neurotransmitters that depolarize a neuron are said to be excitatory, and those that hyperpolarize are inhibitory, referring to their effects on neuron activity.

There are a variety of neurotransmitters, and some neurotransmitters stimulate neuron activity while others inhibit it, according to the charge of the ion fluxes they stimulate. Otto Loewi, a German-born pharmacologist, discovered the first neurotransmitter, acetylcholine, which can be either excitatory or inhibitory depending on the synapse type. Subsequently, three amino acids were found to also be important transmitters: (1) glutamate, by far the most important neurotransmitter, which is excitatory and controls so-called ionotropic glutamate receptors that for the most part admit sodium ions. Thus, the activity of a synapse employing glutamate as a neurotransmitter will depolarize a neuron, which in turn stimulates neuron activity; (2) GABA (gamma aminobutyric acid), which activates chloride channels and hyperpolarizes neurons due to the chloride ion's negative charge and its flow into the neuron. GABA appears to always inhibit neurons; (3) glycine, which binds an ionotropic receptor that conducts chloride and is inhibitory. Thus, neurotransmitters can have either a local excitatory (depolarization) or a local inhibitory (hyperpolarization) effect on a neuron, depending on the neurotransmitter and the receptor types at the synapse as well as the ion flows the receptors mediate.

Glutamate and GABA also activate so-called metabotropic receptors that belong to the 7-transmembrane, G-protein-coupled receptor family. These receptors most often have a single subunit whose peptide chain passes through the plasma membrane seven times (hence their name). They have no pore for conducting ions but instead induce biochemical signals in the neuron by controlling cytoplasmic signaling proteins when the receptor binds its ligand, glutamate, or GABA.

The Action Potential

Any neuron may have a thousand or more synapses on its dendrites and cell body, each employing a particular neurotransmitter. From these thousands of inputs, the postsynaptic neuron sums the positive and the negative influences. Glutamate is released at synapses with glutamate receptors that conduct Na^+ ions into the postsynaptic neurons. If by this means glutamate depolarizes a neuron sufficiently and the electrical potential (voltage) of the interior of the neuron reaches a value more positive than a certain value, the neuron is said to reach "threshold," and this triggers an action potential, an electrochemical event often referred to as "firing."

During an action potential, sodium ions enter the axon through a second type of channel called a voltage-gated sodium channel. This entry depolarizes adjacent positions in the axon, and because voltage-gated sodium channels lie in a series, arranged down the length of the axon, the opening of one opens the neighboring voltage-gated sodium channels. This forms an electrophysiological wave of depolarization that moves down the length of the axon to the axon terminal. The arrival of the action potential wave at the axon terminal induces the release of a neurotransmitter, which crosses the synaptic cleft to stimulate the next neuron in the circuit. Hodgkin and Huxley, using a fine capillary threaded down a giant squid axon as an electrode, were the first to record an intracellular action potential.

Some neurons have evolved a mechanism for speeding up the action potential as it moves down the axon. This mechanism employs myelin, a kind of electrical insulation that is laid down on axons by Schwann cells in patches one to two millimeters long. In between the myelin sheath patches, there are exposed areas of the axon called nodes of Ranvier, which is where the sodium channels are located, and the action potential jumps rapidly from node to node, increasing the speed of transmission.

The transmission of signals between neurons at synapses enables the brain to function as what is perhaps the most highly organized information-processing system in nature. If we consider that even a single synapse is subject to many forms of regulation, and that an individual neuron can possess over a thousand synapses, and that there are about one hundred billion neurons in the human brain, we can begin to appreciate the exquisite level of information encoding and processing that the brain may achieve. But synapses and neurons alone, no matter how many, cannot serve us unless they are organized anatomically into a functional tissue. For this purpose, genetically programmed, developmental processes establish the cortical sheets and specialized brain nuclei that form the brain's highly complex but precisely organized network of circuits. Genetically programmed neurodevelopment in humans continues until adolescence, when environmental forces arising from personal experience take over regulation of the function of the brain circuitry.